PRAISE FOR GENE MACHINE

'An enchanting and invigorating work, *Gene Machine* casts a many-angled light on the world of science, the nature of discovery, and on one of the deepest mysteries of twentieth-century biology. Ramakrishnan, one of the key players in deciphering the molecular basis of protein translation, gives us both a rollicking scientific story and a profoundly human tale. In the tradition of *The Double Helix*, *Gene Machine* does not hesitate to highlight the *process* by which science advances: moving through fits and starts, often underscored by deep rivalries and contests, occasionally pitching towards error and misconception, but ultimately advancing towards profound and powerful truths. An outsider to the world of ribosome biology – an Indian immigrant, a physicist by training – Ramakrishnan retains his "outsider's" vision throughout the text, reminding us about the corrosive nature of scientific prizes, and the intensity of competition that drives researchers (both ideas, I suspect, will have a munificent effect on our current scientific culture). Ramakrishnan's writing is so honest, lucid and engaging that I could not put this book down until I had read to the very end.'

Siddhartha Mukherjee,
author of *The Emperor of All Maladies* and *The Gene*

'Discovering the structure of the ribosome was a truly incredible moment in the history of humankind: this intricate, microscopic machine that lies at the heart of all life, made mostly of RNA, that mysterious material that pre-dates both DNA and protein. As its shape and moving parts came gradually into focus through ingenious applications of crystallography, it is extraordinary to think that this is a device vital to all life, yet which no living thing has seen or understood till now. In this detective story of a book, told with smiles and subtlety, Venki Ramakrishnan relates how he, an immigrant from India, managed to assemble the people, the ideas and the tools to achieve this remarkable feat, in collaboration and (sometimes sharp) competition with other scientific teams, culminating in a Nobel Prize. For students of how science actually happens, this is a book to be treasured and pored over.'

Matt Ridley,
author of *Genome* and *Francis Crick: Discoverer of the Genetic Code*

'The ribosome is the central processor that decodes the universal machine-code of life, and the history of its unravelling is on a par with that of DNA itself. You could think of Venki Ramakrishnan as a sort of "nice Jim Watson". His meticulously detailed and generous memoir has the same disarming frankness as *The Double Helix*. His personal honesty about the competitive ambition that drove him is tempered by his deeply thoughtful reflections on the potentially corrupting effect of big prizes. *Gene Machine* will be read and re-read as an important document in the history of science.'

Richard Dawkins,
English ethologist, evolutionary biologist and author

'This exhilarating account of the race to understand the molecular machine that turns genes into flesh and blood is remarkable for its candid insights into the way science is really done, by human beings with all their talents and foibles. Venki Ramakrishnan, an outsider in the race, gives an insider's view of the decades-long quest to map the million atoms in the machine to fathom the fundamentals of life, pave the way for new antibiotics, and share the glory of the Nobel Prize.'

Roger Highfield,
Director of External Affairs, Science Museum Group

'Quite a ride. This is a riveting personal account of the race to decipher the structure of the ribosome, one of the most complex and fundamental machines in the cell. The book takes up the baton from Watson's *The Double Helix*, and like Watson, Ramakrishnan is disarmingly candid in tone, sometimes disquietingly so. His telling is laced with wisdom spun from a remarkable life story and the sharp lab anecdotes that are the lifeblood of everyday science.'

Nick Lane,
author of *The Vital Question*

'The ribosome, a structure of astonishing complexity, "lies at the crossroads of life" and Venki Ramakrishnan played a key role in revealing its biological mysteries. His superb account lays out the science with great lucidity, but he also grants us the human face of science – the hard work and brilliant insights, of course, but also the role of luck, of personalities, jealousy, money, the roulette of major awards, and the further rewards heaped upon the fortunate. Science, in his glorious telling, becomes "a play, with good and bad characters". Competition and collaboration can appear inseparable, crucial figures get overlooked. It's a wonderful book and a great corrective to the notion of science as dispassionate, untainted objectivity.'

Ian McEwan,
author of *Atonement*

'If someone had told me that one of the most witty and enthralling books I'd read this year would be on the quest to understand ribosomes, I believe I would have laughed in his face, but I would have been quite wrong. *Gene Machine* is beyond superb.'

Bill Bryson
author of *A Short History of Nearly Everything*

'This book is dynamite. Like no science book ever before, this is an honest, frank and simply jaw-dropping account of how a relative outsider ended up winning a Nobel Prize.'

Daniel M. Davis,
author of *The Beautiful Cure*

'In *Gene Machine* one of the world's leading scientists reveals the reality of scientific discovery and the rivalry, collaboration and thrills that are involved. The result is a brilliant under-the-hood account of what it takes to win the Nobel Prize. Exciting and brutally honest, Venki's book explains the dramatic turns in the race to describe the structure of the ribosome – an essential component of every cell that has ever lived. I laughed out loud, I shouted in disbelief, and I learned so much from reading this book.'

Matthew Cobb,
author of *Life's Greatest Secret*

'A must-read for anyone interested in a glimpse of the messy business rivalries, failed experiments, the frustration of mistakes behind how science happens.'

The Times

'An engaging and witty memoir … that highlights how science actually works… This profoundly human story is written with honesty and humility… Anyone who is captivated by an absorbing story well told will find much to appreciate in this fascinating book.'

Forbes

'This is not an objective history of the field, but a highly personal account. As such, anyone who wants to know how modern science really works should read it. It's all here: the ambition, jealousy and factionalism – as well as the heroic late nights, crippling anxiety and disastrous mistakes – that underlie the apparently serene and objective surface represented by the published record.'

Nature

'It is [Ramakrishnan's] full embrace of the role of the antihero that makes *Gene Machine* so much fun to read and also serves as a reminder to us all of the beating human heart that lies at the center of every advance in science.'

The Wall Street Journal

'[O]ne of the most valiant accounts of the triumph of instruments and methods ... The book offers invaluable, readable insights to the aspirant scientist.'

The Hindu

'Ramakrishnan has written a deeply absorbing human story of how modern science is conducted, which is no longer a solo exercise but involves big teams of equally competent individuals – a far cry from the days of Newton and even Einstein when scientists worked in relative isolation.'

C.P. Rajendran,
Open magazine

'*Gene Machine* is about an obscure subdomain of biochemistry, but it is told like a Promethean hero's journey. Like all epics, it is riveting...'

The Indian Express

'An enduring autobiographical classic … it offers a beautiful window into the life and thoughts of a great scientist.'

Business Standard

GENE MACHINE

The Race to Decipher the Secrets of the Ribosome

VENKI RAMAKRISHNAN

HarperCollins *Publishers* India

First published in hardback in India in 2018 by
HarperCollins *Publishers*

This edition published in paperback in India in 2020 by
HarperCollins *Publishers*
Building 10, Tower A, 4th Floor, DLF Cyber City, Phase II,
Gurugram Haryana – 122002, India
www.harpercollins.co.in

First published in Great Britain and Australia
by Oneworld Publications, 2018

8 10 9

This Licensed Edition of *Gene Machine* is published by
arrangement with Oneworld Publications.

P-ISBN: 978-93-5357-471-0
E-ISBN: 978-93-5302-324-9

Typeset by Jeff Williams

Printed and bound at
Nutech Print Services - India

FSC
www.fsc.org
MIX
Paper | Supporting responsible forestry
FSC® C125953

This book is printed on FSC® certified paper
which ensures responsible forest management.

Dedicated to Graeme Mitchison
(1944–2018)

CONTENTS

FOREWORD

BY JENNIFER DOUDNA

THIS IS A PERSONAL STORY of the author's experiences as a student, professor, and avid experimentalist seeking to understand how cells carry out one of the most ancient and fundamental of activities, the synthesis of proteins. The passion for discovery, the frustration of experiments gone awry, the personal and professional struggles that accompanied the road to scientific success come alive in this engaging account.

The author's perspectives are unique in several ways. As an immigrant to America and later to England, and as a physicist entering the world of biology, the story captures the feelings of an outsider yearning to be part of the scientific and social world he observes. This yearning may have played a role in the approach to research that he describes: there is at once a desire to belong and yet a willingness to be a maverick and to embark on a journey of discovery that seemed at the outset to be a long shot. Then there is the scientific discovery itself: revelatory structures of the very machinery that reads the genetic code and translates its nucleic acid sequence into chains of amino acids that form all of the proteins necessary for life on earth. The ribosome, comprised of one large and one small subunit, is uncovered in all its glory, with the author's own work revealing the molecular underpinnings of the decoding mechanism of the ribosomal small subunit and the means by which multiple antibiotic drugs block the action of the bacterial ribosome and thereby eliminate microbial infections. The process of determining structures of the ribosomal small subunit, from initial efforts with individual components to later tour-de-force attempts to purify and crystallize the intact small subunit, is a fascinating tale of ingenuity, luck, and ultimate success.

The story is also one of professional dilemmas, the serendipity of discovery, and the deeply human nature of research, in which personalities play an essential role. Any major scientific breakthrough involves multiple contributors as well as the interactions of scientists grappling with the challenges of discovery and setback that accompany the winding path to understanding. It is not always clear how ideas emerge, whether they are truly one's own or instead an outcrop of discussions with others. Sometimes competition comes between colleagues and must be managed even as the joy of discovery is at hand. The award of the 2009 Nobel Prize in Chemistry gave recognition to the early success with ribosome crystallization by Ada Yonath and the subsequent achievements of Venki Ramakrishnan and Tom Steitz in solving the structures of the ribosomal subunits. Yet, as described in the book, others who made key contributions were, by the three-recipient limitation of the prize, left out of this recognition despite the importance of their insights. The book provides a readable account of the author's experiences and views on these matters that must be taken in the spirit of a memoir rather than an objective historical essay. Students of science and of the scientific method will find this story to be a fresh take on the process of discovery and the sometimes tortuous trail that leads to new knowledge. In the end, it is a fascinating contribution to the scientific literature and one that will be valued as both a description of fact and a dissection of the emotional side of a scientist's approach and eventual achievement.

J. A. Doudna

PROLOGUE

LOOKING BACK, IT IS SURPRISING how little effect her visit had. It was a grey autumn day in 1980. A small advertisement on a Yale University noticeboard had announced a lecture with a vague title. I had no trouble finding a seat, even though I arrived just before the speaker, because only a few specialists had bothered to come.

She strode in exuding confidence, even audacity. After a brief introduction by her host, she began to describe the efforts of her group in Berlin to obtain crystals of an enormous assembly of molecules involved in translating genes into proteins. At the time, obtaining crystals was a key step to deciphering their structure.

After her talk, there were hardly any questions because we weren't sure what to make of the work. We thought it was astonishing that someone had coaxed such a large and floppy particle to form the regular three-dimensional stacks of molecules that make up a crystal. As we walked down the hall back to our labs, a colleague of mine teased another, saying, 'How come you can't even crystallize a tiny piece of it and she's done the whole thing?' However, her crystals were not good enough to obtain a structure, and at the time, nobody even knew how to figure out the structure for something so large. So in the end, we thought it was an interesting curiosity, but none of us felt that our world had changed and we needed to drop what we were doing.

Little did I know that the scientist, Ada Yonath, would figure in my professional life for the next three decades; that I would compete with her and others in an intense race to understand an object that is at the heart of all life; or that one December, I would be sitting between her and the crown princess of Sweden at a Nobel Prize banquet in Stockholm.

CHAPTER 1

An Unexpected Change of Plans in America

WHEN I LEFT INDIA, I had my heart set on becoming a theoretical physicist. I was nineteen years old and had just graduated from Baroda University. It was customary to stay on to get a master's degree in India before going abroad for a PhD, but I was eager to go to America as soon as I could. To me, it was not only the land of opportunity but also of rational heroes like Richard Feynman, whose famous *Lectures on Physics* had been part of my undergraduate curriculum. Besides, my parents were already there, as my father was taking a short sabbatical at the University of Illinois in Urbana.

Since this was a last-minute decision, I had not taken the GRE exam that is a requirement for American graduate schools, and most universities would not even consider my application. The physics department at the University of Illinois initially accepted me, but when the graduate college found out I was only nineteen, they said that at best I could join as an undergraduate with two years of college credit. No middle-class Indian then could afford the cost of tuition and living in America. In the meantime, my department chairman in Baroda showed me a letter from Ohio

University asking him to let prospective graduate students know about its programme. I had never heard of Ohio University before but saw that the department had an IBM System/360 computer and a Van de Graaff accelerator and that its faculty had been trained at some of the best universities; that seemed good enough for me. They waived the usual GRE requirement and accepted me with financial support. After the typically nerve-racking interview for a student visa at the US consulate in Bombay, I bought my ticket to the promised land.

As soon as I had finished my final exams, I left the sweltering heat of India and set off for America. I had caught a fever and the flight seemed interminable, stopping at Beirut, Geneva, Paris, and London before landing in New York. I boarded a plane for Chicago and then took a short flight to Champaign-Urbana. As I stepped out onto the tarmac on the evening of May 17, 1971, I felt a blast of the coldest wind I had ever experienced.

My sudden immersion into American college life came as something of a shock. College life in India was fairly staid. Students were conservative in their attire and focused on their studies; many, like me, still lived with their parents. Dating, and especially premarital sex, was quite unusual. I arrived as a geeky student with a crew cut, glasses with thick black plastic frames, and orange suede shoes two sizes too large, into an America that in 1971 was a continuation of the sixties. American students seemed to belong to an entirely different species: the males in tattered jeans and hair even longer than the females, who in their hot pants and halter tops seemed almost naked compared to the Indian women I had just left behind. Campuses throughout America were protesting against the Vietnam War. One afternoon, out of a mixture of curiosity and sympathy, I attended a peace rally. I stuck out like a sore thumb but then found two older guys at the back, who like me had short hair and were dressed in the same cheap polyester trousers and shirts. I walked over to them and tried to be friendly, but they were rather curt and suspicious. Only later did I find out that they were FBI agents keeping an eye on the anti-war troublemakers.

I spent the summer taking courses at the University of Illinois, filling gaps in my education from Baroda. At the end of the summer, I drove with my parents and sister to the pretty and hilly university town of Athens in southern Ohio, which was to be my home for the next few years. The first problem was finding any home at all. Since I had to live on my teaching assistantship and was a vegetarian, I thought it would be best if I rented a small flat where I could cook my own meals. We scoured the newspaper for rental ads with little success. In one case, the landlady said the flat was available, but when we showed up to look at it only a few minutes later, she took one look at me and said it was 'just taken.' That was my first experience of racism in America. Unable to find a flat that weekend, I signed up for a dorm room and spent the first year living primarily on cheese sandwiches at the cafeteria.

Notwithstanding its culinary disadvantages, the dorm room had the great benefit of allowing me to instantly acquire a group of friends and avoid the feeling of isolation and ghettoization so common to foreigners. My dorm mates quickly helped me assimilate into American college life. The first Saturday, we went to a football game. The pomp, with cheerleaders, bands, and the loud PA system, seemed to dwarf the game itself.

The dorm also had the advantage of being close to the physics department, and several fellow graduate students lived in rooms nearby, so we were able to form a friendly study group and get used to graduate school together. Physics graduate students generally have to do a year or two of coursework followed by a comprehensive exam before they begin serious research. Although I finished my coursework and the written part of the comprehensive exam without too many problems, the oral part at the end gave me the first inkling that I did not have a burning desire to be a physicist. I was asked what recent interesting discoveries in physics I had read about. I couldn't name even one, and only after some prodding could I even name an area I found interesting. They passed me anyway, and I decided to work under the supervision of Tomoyasu Tanaka, a well-regarded condensed matter theorist. By then, I had

Figure 1.1 The author as a graduate student in physics at Ohio University

already become intrigued by biological questions, and I included some biological problems in my thesis proposal. Since neither Tomoyasu nor I knew the slightest thing about biology, these proposals were pure fantasy and I soon abandoned them.

As I began my thesis work, I realized that I could not quite see how to identify key questions, let alone how to approach them. Even worse, I didn't find my work interesting. I retreated into my social life, playing on the university chess team, going hiking with my friend Sudhir Kaicker, learning about Western classical music from another friend, Tony Grimaldi, and generally doing anything except making progress on my graduate work. Tomoyasu was an almost stereotypically polite Japanese who would occasionally come into my office to delicately enquire about my progress, and I would tell him in a roundabout way that there hadn't been any. This continued for a couple of years. I've often said that if I'd had students like me, I would have fired them!

Things suddenly changed when I met Vera Rosenberry, a recently separated woman with a four-year-old daughter. Some mutual friends decided that the two of us should meet, perhaps because we were both vegetarians – an oddity in 1970s southern Ohio. I was completely oblivious that our first meeting was a setup because we were part of a large Thanksgiving gathering. Observing my cluelessness, my friends decided I needed additional help and invited me to a dinner party with just one other couple. Vera struck me as both intelligent and classically good looking, but I assumed someone like her would be out of my league and couldn't possibly be interested in me. So I tried to introduce her to a friend by inviting him to dinner along with Vera and her daughter, Tanya. I spent some of that time playing with Tanya so that my friend and Vera could be free to chat. It was my friend who had to point out that she seemed interested in me not him, and if anything, she probably became even more interested when she saw how well I got along with her daughter. Despite this comically inept start on my part, we began a stormy courtship that lasted less than a year and got married soon after her divorce became final. At the age of twenty-three, I found myself married and the stepfather of a five-year-old girl.

Marriage, however, focused my mind on my career. Vera wanted to have another child, so I faced the prospect of supporting a family without any idea of what I would do next. There seemed no question that if I stayed on in physics, I would spend the rest of my life doing boring and incremental calculations that wouldn't result in any real advance in understanding. Biology, on the other hand, was undergoing the sort of dramatic transformation that physics had in the early twentieth century. The revolution in molecular biology that began with the structure of DNA was still going strong, and we were beginning to get fundamental insights into the molecular basis of biological processes that had puzzled us for centuries. Almost every single issue of *Scientific American* would report some major breakthrough in biology, and it seemed as if it could be done by mere mortals like me. My problem was that I knew only the most basic biology and had no idea what biological research entailed. So even before I had finished my physics PhD, I made the difficult decision to attend graduate school all over again, this time in biology, taking heart from the fact that many illustrious scientists like Max Perutz, Francis Crick, and Max Delbrück had made a similar transition.

I wrote to several top-tier universities, but many of them did not want to accept someone who already had a PhD into graduate school. Two replies were particularly memorable. The first, from Franklin Hutchinson at Yale, was a friendly letter saying that although they couldn't accept me as a graduate student, he would send around my CV to his faculty in case someone was interested in hiring me as a postdoctoral research fellow. Two of them wrote back: Don Engelman and, ironically in hindsight, Tom Steitz. I thanked them both and said that I didn't have enough background to be useful to them as a postdoc and would try to acquire some training first. At the opposite end of the spectrum from Hutchinson was James Bonner of Caltech. In my applications, I had written that because I was only twenty-three, I was young enough to go to graduate school again. Bonner berated me for bragging about my age, adding that he, too, was only twenty-three when he received his PhD and was considered backward in his family. He also said that the areas I had mentioned – allostery, membrane proteins, and neurobiology – were hardly surprising because they were the most fashionable areas of biology. If I wanted to work in them, he wrote, I had to first show that I could actually be competent in those areas, and certainly Caltech would not accept me as a student. Perhaps he had never read *Catch-22*. Fortunately for me, Dan Lindsley of UC San Diego was willing to accept me into the biology department as a graduate student with a fellowship. Even more fortunately, Vera and Tanya were game to move to California and continue living on a meagre graduate student's stipend with the added burden of an infant – all without a car.

Somehow, I scraped together enough work to produce a passable thesis just in time. Our son Raman was born just a month after my PhD exam. A couple of weeks later, a friend and I drove from Ohio to California in a Ryder truck with all our belongings, and Vera and the children flew in with my mother-in-law a week later. Once we had settled in, I began my studies in earnest in the autumn of 1976.

The thing that immediately struck me about biology was how many facts you have to know. The introductory lectures for new graduate students were full of jargon I didn't understand at all. To catch up, I took a full load of undergraduate courses in genetics, biochemistry, and cell biology, in addition to doing first-year graduate student rotations, which are short six-week projects that American grad students typically do before they settle into a lab for their PhD research. Because my physics research had been entirely theoretical, I was totally ignorant of laboratory work. This was brought home to me during a rotation in Milton Saier's lab, which worked on sugar uptake by bacteria. The experiment entailed adding a certain amount of radioactive glucose to a culture of bacteria at time zero, and then measuring how much glucose had entered the bacteria at various time points. The amount of the glucose to be added was much smaller than anything I had encountered before – only about twenty microlitres (less than 1 per cent of the volume of a teaspoon). How could you measure such a small volume? I asked. The technician who was training me very pleasantly showed me a device called a Pipetman, which is essentially a tube with a piston that can be set so that it can go up or down by a precise amount. She showed me how to set the volume on the dial, how to draw up the right amount, and how to give the knob a little extra push at the end to make sure all the sample had been ejected. That was all there was to it, she said. I took the device and plunged it into the radioactive glucose, and she exclaimed, 'What the hell are you doing? You have to use *tips!*' The devices were so commonplace that she'd forgotten to mention the narrow disposable plastic tips that needed to be attached to the end of the Pipetman so that it never got contaminated by contact with the sample.

Moving with a young child and an infant was not exactly conducive to learning a new field. However, I was extremely lucky that Vera, who was starting her own career as a children's book illustrator, could work from home. She did almost all of the childcare and housework, allowing me to focus on my studies. I ended the first year feeling optimistic that I had a broad enough background in biology and fairly varied lab experience. In my second year, I began work with Mauricio Montal, who was studying proteins that let ions pass through the thin membranes of lipids that envelop all cells. As it turned out, I wouldn't stay in his lab very long. Almost entirely by chance, I would move across the country yet again to begin work on one of the oldest and most central molecules of life.

CHAPTER 2

Stumbling into the Ribosome

MENTION DNA AND ALMOST EVERYONE nods in understanding. We all know – or think we know – what DNA means. It determines the essence of who we are and what we pass on to our children. DNA has become a metaphor for the fundamental qualities of almost anything. 'It is not in their DNA,' we say, even when referring to a corporation.

But mention the word *ribosome* and you will usually be met with a blank stare, even by most scientists. A few years ago, I was told by Quentin Cooper on the BBC radio programme *Material World* that the previous week's guest had been outraged that the eye only merited half a programme when an entire episode was planned for a mere molecule like the ribosome. Of course, not only are most of the components of the eye made by the ribosome, but virtually every molecule in every cell in every form of life is either made by the ribosome or made by enzymes that are themselves made by the ribosome. In fact, by the time you read this page, the ribosomes in each of the trillions of cells in your body will have churned out thousands of proteins. Millions of life forms exist without eyes, but every one of them needs ribosomes. The discovery of the ribosome and its role in making proteins is the culmination of one of the great triumphs of modern biology.

When I arrived in California to learn biology, like most physicists, I had no idea what the ribosome was and only the vaguest idea of what a gene was. I knew genes carried the traits we inherited from our ancestors and passed on to our offspring. But I learned that genes are much more. They are the units of information that allow a whole organism to develop from a single cell like a fertilized egg. Although nearly all cells contain a full set of genes, different sets of genes are turned on or off in different tissues, so a hair or skin cell is very different from a liver or brain cell. But what actually are genes?

Broadly speaking, a gene is a stretch of DNA that contains information on how and when to make a protein. Proteins carry out thousands of functions in life. For example, they are what make muscles move. They let us sense light, touch, and heat and help us fight off diseases. They carry oxygen from our lungs to our muscles. Even thinking and remembering is made possible by proteins. Many proteins called enzymes catalyse the chemical reactions that make the thousands of other molecules in the cell. So ultimately, proteins not only give a cell its structure and shape but also enable it to function.

Understanding how the information in a piece of DNA could be used to make a protein was the culmination of an exciting decade that began with the classic 1953 paper on the double-helical structure of DNA by James Watson and Francis Crick. Often, the structure of a molecule does not immediately explain how it works. Not so with DNA, which immediately suggested both how it could carry information and how it could reproduce itself. It had long been a mystery how information in a cell is duplicated when it divides or how offspring inherit this information when an organism reproduces.

In each molecule, the two strands of DNA that intertwine to form a double helix run in opposite directions. Each strand has a backbone of alternating sugar and phosphate groups, and one of

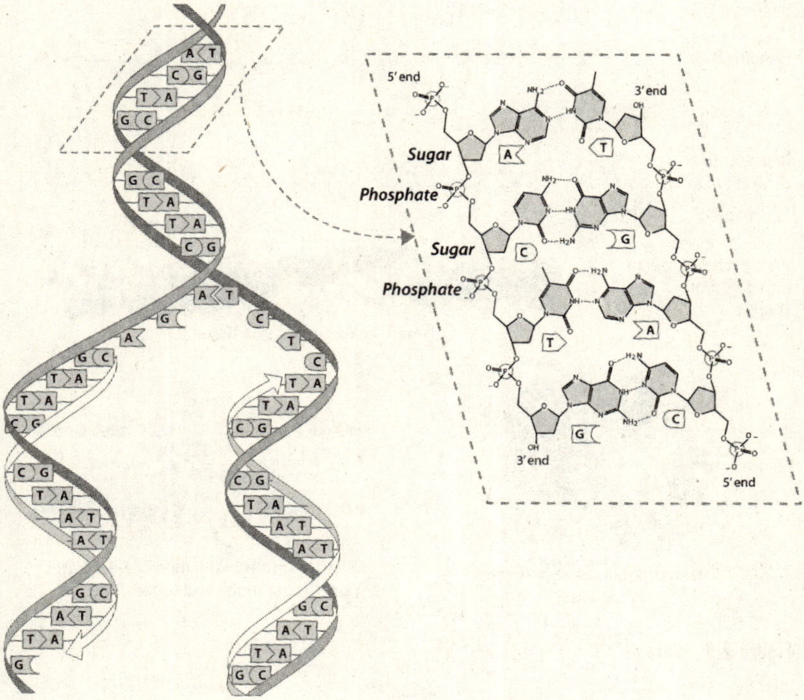

Figure 2.1 DNA structure

four types of bases – A, T, C, or G – is attached to the sugar and faces the inside of the helix. While playing with cardboard cutouts of the bases, Watson arrived at a brilliant insight: he realized that an A on one strand could chemically bond or pair to a T on the other but not to any of the other bases, while a G on one strand could similarly pair with a C on the other. In doing so, the shape of each base pair, whether it was AT or CG, was about the same, which meant that regardless of the order of the bases, the overall shape and dimensions of the double helix was about the same. This formation of *base pairs* meant that the order of the bases on one strand would precisely specify the order on the other strand. When cells divided, the two strands would separate, and each would have the

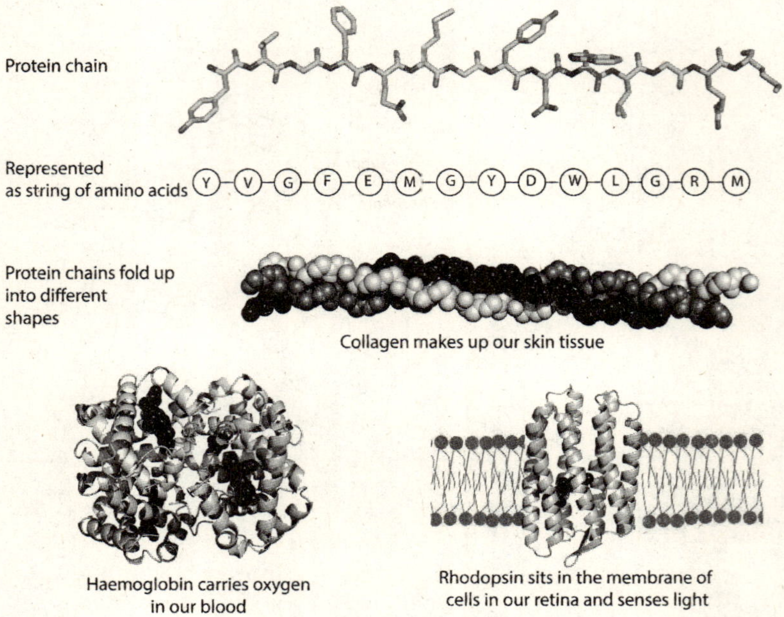

Figure 2.2 Proteins

information to serve as a template to make the other strand, resulting in two copies of the DNA molecule from one. In this way, genes were able to duplicate themselves. After centuries, we finally understood in molecular terms how hereditary traits could be transmitted from generation to generation.

The structure immediately suggested how genes could be duplicated and passed on but not how the information in our genes could actually be used to make proteins. The problem was that each strand of DNA was a long chain made up of building blocks containing the four types of bases. But proteins are completely different chains made up of amino acids, and their chemical linkage is completely different. Their enormous variety comes from the fact that there are twenty types of amino acids, which have a wide range of chemical properties. The length and order of amino acids in each

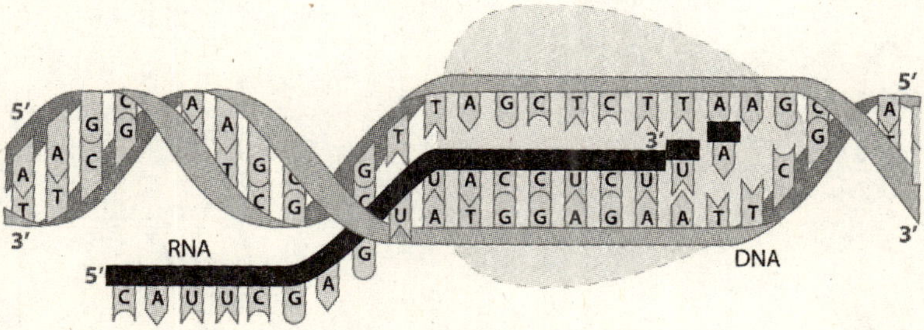

Figure 2.3 Transcription: copying a gene in DNA to messenger RNA

protein chain is unique, and amazingly it contains the information needed for the chain to fold up into its own unique shape, which allows it to carry out its special function. Crick realized that the order of bases in DNA coded for the order of amino acids in a protein, but the question was how.

Lots of people worked on this problem for well over a decade. It turns out that a stretch of DNA containing a gene is copied into a related molecule called messenger RNA or mRNA, so-called because the molecule carries the genetic 'message' to where it is needed. RNA, which stands for ribonucleic acid, differs from DNA or **deoxy**ribonucleic acid by having an extra oxygen in the sugar ring. It, too, has four bases, but the base thymine (T) in DNA is replaced by a very similar base, uracil (U), in RNA, which, like T, also base pairs with A.

How do you go from four types of bases to twenty types of amino acids? It would be like following a long sentence of instructions written in code using a foreign alphabet. It turns out that the bases are read in groups of three at a time, and each group is called a codon. The way they are read – something predicted by Crick – is that another RNA molecule called transfer RNA or tRNA has an appropriate amino acid attached at one end and a group of three bases called an anticodon at the other end. The anticodon and codon form base pairs just like the ones between the two strands of

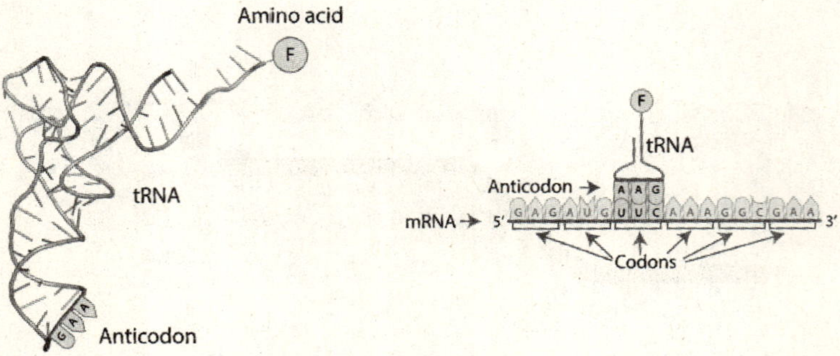

Figure 2.4 Transfer RNA: the adaptor molecules that bring amino acids and read the code on messenger RNA

DNA. The next codon is recognized by a different tRNA, which brings along its amino acid, and so on.

The next big discovery was that this doesn't happen by itself. Cell biologists discovered particles in cells where the mRNA is read and proteins are made. The particles were tiny by normal standards – you could pack four thousand of them in the width of a human hair. There were thousands of them in every cell, from bacteria to humans. But they were enormous in molecular terms. Each of them contained about fifty proteins and three large pieces of their own RNA – a *third* type of RNA (after mRNA and tRNA). Initially, scientists referred to the particles as 'ribonucleoprotein particles of the microsomal fraction' because they were made up of both RNA and protein and were isolated from cellular fragments known as microsomes. This was quite a mouthful to say; so in the late 1950s at a conference, Howard Dintzis suggested the name ribosome, which it has been called ever since. Dintzis was also the first person to figure out the direction in which a protein chain is made. Embarrassingly, even after working for thirty years in the field, I didn't know Dintzis or his work. When I finally met him in 2009 at Johns Hopkins University, where I had been

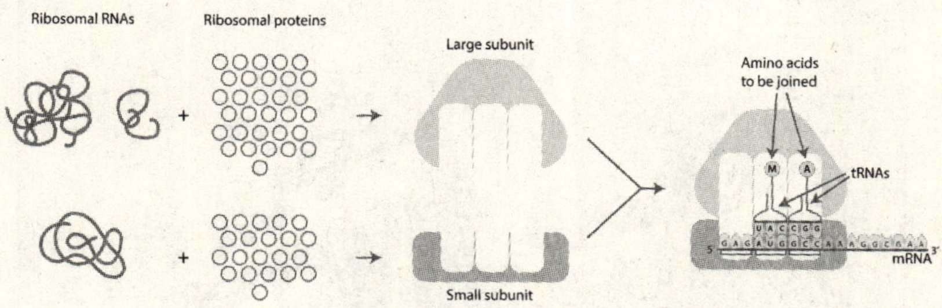

Figure 2.5 Composition of ribosomes

invited to give a lecture named after him, he was still justifiably proud of having coined the word.

The whole ribosome has over a million atoms. Because it is the link between our genes and the proteins they specify, the ribosome lies at the very crossroads of life. Even though everyone knew this, nobody had any idea what ribosomes looked like other than that they were blobs consisting of two parts. And that was a real problem. Somehow, the ribosome bound mRNA and stitched together the amino acids brought by the tRNAs into a protein. But without knowing what it looked like, how could we possibly understand how it all worked?

Imagine you are a Martian peering at earth from above. You observe tiny objects on the surface that move mainly in straight lines, occasionally turning at right angles. If you were able to get a little closer, you might see that these objects move only when even smaller objects enter them and stop moving when they leave. If you have sensors, you could tell that they consume hydrocarbons and oxygen and emit carbon dioxide and water along with some pollutants and heat. But you would have absolutely no idea what these objects really are, let alone how they work. Only by knowing the detailed construction of the object would you be able to see that it is made of hundreds of components that work together and that it has an engine connected to a crankshaft that makes the wheels turn. You

Figure 2.6 Alfred Tissières and James Watson, two early pioneers of ribosome research *(courtesy of Cold Spring Harbor Laboratory)*

would need to see even more detail to know that the engine itself has chambers with pistons and draws in a mixture of fuel and oxygen that is ignited with a spark plug, which drives the piston forward.

It is the same with understanding molecules. Knowing the detailed structure of DNA revolutionized our understanding of how it works to store, transmit, and replicate genetic information. But the ribosome was not a simple molecule like DNA. It was enormous and complex and seemed just too daunting and intractable.

Many great scientists like Crick, who had played key roles in figuring out how information in DNA is encoded, simply gave up on the ribosome and left for other fields. Sydney Brenner, an equally eminent colleague of Crick's who was one of the discoverers of mRNA, said in the 1960s that the structure of the ribosome was a trivial problem and there was no need to work on it in Cambridge since that sort of work would, of course, be done by Americans. This reminds me of Senator George Aiken saying of the intractable Vietnam War that the 'US should declare victory and get out.' One of the early molecular biologists who persisted on the ribosome was Watson, who worked on the problem with Alfred Tissières, a biochemist from Geneva who was visiting his lab. Nearly forty years later, at a meeting in Cold Spring Harbor in 2001, Watson recalled those early days, saying that when he realized how complex the ribosome was, he automatically knew that we would never know its structure.

The ribosome was far from my mind as I settled into Mauricio Montal's lab, but after I had only been there a few months, I came across an article on the ribosome in *Scientific American* that would change my life. It described how to locate the many different proteins on the ribosome using neutron scattering – a technique known to physicists but hardly used in biology. The two authors were Don Engelman and Peter Moore, and I remembered that Don was one of the people who had expressed interest in having me as a postdoc when I was trying to switch from physics to biology. I thought if he'd wanted me with no biological background at all, he might be even more interested now that I had learned some biology and had over a year of lab experience. It also occurred to me that I had already learned enough biology to do research in the field, and there was no need to get a second PhD in biology.

So I wrote to Don reminding him of our previous correspondence, saying I was now more prepared for a postdoc. Since I knew Don's main interest, like Mauricio's, was in membranes and proteins in membranes, I told him I would like to work on them in his lab. He wrote back saying that he didn't have any positions but his collaborator Peter Moore did, and if I came there and worked on ribosomes, I could then do some membrane work in my spare time. By that time, I knew that ribosomes were fundamentally important, so I said that was fine with me. As it turned out, the 'spare time' was non-existent.

Peter soon wrote saying he was visiting San Diego for a conference and would be happy to meet me. When I went downtown to see him, he was dressed in his characteristically preppy clothes with a brown corduroy jacket, and his thick glasses and manner completed the stereotype of an Ivy League intellectual. Which indeed he was. He entered the fast track early in life and never left it, and I was never sure whether he really understood what it was like not to have been in elite institutions all your life. His father had made pioneering contributions to transplantation surgery at Harvard, and Peter himself had studied at a private school and then Yale before going to Harvard for grad school, where he worked with Watson on the ribosome. After that, he went to work with Watson's friend and collaborator Alfred Tissières in Geneva, who by then was a leader in the ribosome field. There he set about purifying the different proteins that made up the ribosome.

Realizing that the key to understanding the ribosome was knowing its structure, and that he needed to acquire skills in structural analysis, he left Geneva to go to the Medical Research Council (MRC) Laboratory of Molecular Biology in Cambridge, England. This lab was the direct descendant of the MRC unit where Watson

Figure 2.7 Peter Moore around 1980 when the author worked in his lab at Yale *(courtesy of Peter Moore)*

and Crick had done their work on DNA and by then had become a mecca for studying the structures of all kinds of biological molecules. Americans referred to the lab as the MRC, as if it was the only one of the many labs that the Medical Research Council supported throughout the country that was worth knowing about. The British called it the MRC-LMB or simply the LMB, as it is currently known.

At the end of his stint at the LMB, Peter returned as a faculty member to his alma mater, Yale, where he has been ever since. He has a dry wit with a reservoir of knowledge that ranges from all aspects of science to history and the classics. Although a shy and reserved man, he lost his natural reticence when it came to science. His lectures were articulate and laced with humour, and generations of Yale students and scientists were exposed to his wrath when they presented sloppy arguments.

At the conference in San Diego where we first met, he was standing by himself, waiting for me while people were milling about him. After a brief hello, we chatted about my background and his project. I wasn't at all sure how this informal interview had gone, but he wrote back soon afterwards to invite me to visit Yale. My visit there was very pleasant. Despite my obvious naïveté, Peter formally offered me a position, and I accepted immediately. I spent the rest of the academic year finishing up work in Mauricio's lab. Finally, at the end of the summer, I left for New Haven, picking up my family en route in Ohio, where they had spent the previous few weeks.

I arrived in Peter's lab in the autumn of 1978 with some trepidation. Faced with actually having to do postdoctoral research at Yale, my earlier confidence evaporated because, despite my two years of grad school in biology, I had very limited experience in actual biological research. A few days after I arrived, Peter and I were walking towards each other in a long passageway in the neo-Gothic Sterling Chemistry Laboratory. As soon as we drew close, he suddenly looked away. I was worried that he'd already regretted hiring me, but his longtime technician Betty Rennie laughed and told me that was just his manner. In any case, he was perfectly nice to me, and a year later, he must have felt I was sufficiently competent that he could leave me alone for an entire year while he went away on sabbatical in Oxford. During his absence, I grew a beard that I kept for almost twenty-five years.

By the time I began working for Peter, some basic facts about the ribosome were already established. All ribosomes have two parts, known as the small and large subunits. The small subunit binds the mRNA containing the genetic information, while the large subunit actually stitches together the amino acids brought in by the tRNAs to make a protein. There are three slots for the tRNAs – one to bring in the new amino acid, one to hold the growing protein chain, and one that is kind of a halfway house before being ejected from the ribosome. During the process, the tRNAs move through the ribosome from one slot to the next, and as they move, they effectively drag the mRNA along with them so that the ribosome is, in effect, moving along the mRNA, helping the tRNAs read one codon after another to make the protein. Each step needs the help of proteins that bind and leave the ribosome at various stages, and each step consumes energy. Because it uses energy and moves during this enormously complicated process, the ribosome is referred to as a molecular machine or a nanomachine.

Apart from its fundamental role at the crossroads of biology between genes and the proteins they specify, there was a practical reason to be interested in the ribosome. Over the years, people had realized that many antibiotics work by blocking the ribosome at different steps. Because the ribosomes of humans are sufficiently different from those of bacteria, some of these antibiotics preferentially bind to bacterial ribosomes and are useful in treating infectious diseases. Bacteria are increasingly becoming resistant to infections, however, and knowing exactly how an antibiotic binds to the ribosome could lead to designing new and better ones.

These basic facts had already made it into textbooks, so when I told people I was working on the ribosome, I would often be asked, 'But isn't that already done?' Sometimes this would be accompanied by a pitying look as though I was some poor guy dotting the i's and crossing the t's on a problem that was no longer interesting. The reality was that although we had an outline of what a ribosome did, we had no idea how it did even one of the many complicated steps involved in making a protein. It was as if we knew a bit more about a car – realized it had four wheels and windows and a driver who sat at a steering wheel – but knew nothing else about how it actually worked.

Like many other fields, science has its fashions, and at any given time, some areas are considered more interesting than others. Often these are new areas where people are making rapid advances. Many scientists move on from a problem as soon as it gets too hard to make further progress. Those who are very creative open up entirely new areas, but others just follow one fashionable area after another. If everyone did this, our understanding of phenomena would be quite superficial, but fortunately there are other scientists who stick with a problem, no matter how old and difficult it is, to get right to the bottom of things.

Even though the ribosome had been studied for a couple of decades, nobody even knew where all the fifty or so proteins in it were located, let alone what they did. Peter was collaborating with Don Engelman to tackle this problem. In some ways they could not have been more different. In contrast to Peter's reserve, Don was a tall, gregarious California native with a carefully manicured beard and a booming baritone voice and suave manner that conveyed great authority regardless of the topic of conversation. He had gone to Reed College in Portland, done his PhD at Yale, and then gone to do a postdoc with Maurice Wilkins, the 'third man' of DNA, where he worked on the structure of the membranes that envelope all cells. Unlike Peter, who had spent his entire life working on one or other aspect of the ribosome, Don's interests were more varied.

Don and Peter heard a talk by Benno Schoenborn of Brookhaven National Lab about how neutrons could be used to study biological structures. Neutrons were something only physicists bothered with. Moreover, you needed a nuclear reactor to produce enough neutrons to do an experiment. But the interesting thing about neutrons for biology was that hydrogen and its heavier isotope deuterium interact very differently with neutrons, and hydrogen makes up half the atoms in biological molecules like proteins and RNA.

The talk gave Don and Peter an idea for trying to figure out where the ribosomal proteins were. They realized that if you could somehow make a ribosome in which just two of the proteins had deuterium instead of hydrogen atoms, those two proteins would scatter neutrons differently.

You could get deuterated proteins by growing bacteria in heavy water, which is deuterium oxide. Then you would have to reassemble a ribosome in which any two proteins of your choice were deuterated. Masayasu Nomura at Wisconsin had shown that you could biochemically extract the proteins from the small ribosomal subunit and separate each of the twenty proteins from the mixture by chromatography. You could then mix all the components together in a solution, and under the right conditions, you could reassemble a functional small subunit from the purified proteins and RNA. This meant that you could make a small subunit in which just two of the proteins were replaced by their deuterated counterparts. These ribosomal subunits could then be taken to a nuclear reactor in Brookhaven National Lab in the middle of Long Island and exposed to a neutron beam. Each such experiment would give you a distance between a pair of proteins. By measuring distances between lots of different pairs, you could figure out how they were arranged in three dimensions, very much like how early surveyors mapped unknown terrain by triangulation. The project involved tediously making the same kinds of measurements over and over again with different pairs of proteins in the ribosome.

I joined the lab when only the first few of the proteins had been located this way, and the baton was handed to me by the previous postdoc, Dan Schindler. To my surprise, I learned that neutron beams, even from a nuclear reactor, are orders of magnitude weaker than X-ray beams, so it would take several days to measure the small signal from the deuterated proteins that was buried in the large overall background scattering from the rest of the ribosome. Doing this work during the summer had advantages; while the data were being collected, I would occasionally go to the beach on Fire Island a few miles directly south. At other times, being trapped in Brookhaven was not exactly thrilling because the lab was in an old army camp in the middle of nowhere outside Yaphank. The scientists who worked there lived in communities miles away that were a mixture of rural hamlets and extended suburban sprawl. Unlike a university town with a thriving cultural scene and nightlife, the lab was deserted on evenings and weekends with nothing for a temporary visitor to do. This situation reminded me of a famous *New Yorker* cartoon of the Long Island Expressway that said, 'Exit 66 – Yaphank. If you have already been to Yaphank, please disregard this exit.'

After about three years, a little over half of the proteins were placed in the small subunit, and we wrote a couple of papers on their location. I was wondering how long it would take me to do the remaining ones, but as I approached the end of my fellowship, Don came to me and said it was in my own interest that I move on to the next phase of my career now that I had acquired the training I needed as a postdoc. The project was eventually finished by my successor, Malcolm Capel. The final paper describing the locations of all the proteins showed them as billiard balls superimposed on the shape of the small subunit, and I would joke that about a third of those balls were mine.

Responding to Don's hint, I applied for almost fifty faculty positions, but it was not a great time to be looking for jobs. We had just entered the Reagan era of minimal government, and research funding was tight. Biotech was still in its infancy, and faculty jobs were scarce.

I applied for every opening from junior colleges to universities of varying quality. The smaller teaching colleges looked at my long Indian name and probably worried I couldn't speak English well enough to teach. The universities looked at my career – a bachelor's and doctorate in physics, neither of which was from a prestigious university, two years studying biology without a degree, followed by research using a technique nobody had heard of to work on an old problem that was already unfashionable. No wonder I didn't get a single interview.

Luckily, Oak Ridge National Lab in Tennessee had just started a neutron-scattering facility and was looking for someone to collaborate with biologists, so Don called up the head of the facility, Wally Koehler, to recommend me. In the excitement and optimism of starting my first 'real' job, Vera and I bought a house there right away. In February of 1982, we packed our stuff into our small Ford Fiesta and drove from New Haven to Tennessee through an ice storm in Pennsylvania.

I had gone there on the understanding that I would be able to do my own research, but when I arrived, the biology lab I was promised didn't materialize. Wally Koehler told me that I was there to collaborate with biologists on neutron scattering, not to do my own research. He was a well-known physicist whom I liked personally, but I felt he did not understand the way biology is done and the rather peripheral role neutrons play in biology, and I started looking to leave Oak Ridge very soon after I got there. Luckily for me, Benno Schoenborn, who was developing the biological use of neutrons at Brookhaven National Lab and who had inspired Peter and Don to work on the ribosome, came to the rescue. He offered me an independent job at Brookhaven, which given my situation in Oak Ridge, I was grateful to accept. So only fifteen months after our arrival in Oak Ridge, we sold our house for a substantial loss in the summer of 1983 and moved back to the East Coast, this time to Long Island.

Vera was not happy at leaving her beautiful garden and idyllic life in Oak Ridge, and her heart sank when we drove over the George Washington Bridge and saw all the traffic on the Long Island Expressway. We eventually found a house in East Patchogue, right next to the village of Bellport on the south shore of Long Island. It was a twelve-mile commute each way to the lab and seemed much longer during the ice storms every winter.

Unlike my disastrous experience in Oak Ridge, Brookhaven gave me a well-equipped lab, a technician, and freedom to carry out my own research. My colleagues were very friendly and helpful but made it clear that I shouldn't expect to get tenure for merely continuing work I had done as a postdoc. Fortunately, as a result of collaborations during my brief stint in Oak Ridge, I had developed an interest in chromatin, the complex of DNA and histone proteins that makes up chromosomes in cells. So I began work on how chromatin was organized and for a long time was better known for this than for my continuing ribosome work.

I still kept plugging away at the ribosome using the techniques I had learned, like neutron scattering, but neither I nor anyone else in the field made real progress in understanding how the ribosome actually worked. Individual components of the ribosome seemed to do little by themselves. It was a bit like looking at a set of tyres and pistons in isolation with no idea of how they were put together to make a car. On the other hand, the whole ribosome seemed too large and intractable a problem on which to make any headway. Not only was the ribosome becoming even less fashionable than when I'd started working on it, but neutron scattering was proving a dead end in tackling it or chromatin. Almost a decade after I had switched from physics to biology, it looked like my second career was going down the tubes just like my first one.

CHAPTER 3

Seeing the Invisible

WE SAY SEEING IS BELIEVING, and it is astonishing how often just being able to see things has changed our understanding of the world. For centuries, we had many misconceptions about our own bodies because our knowledge of them came from the Greek physician Galen, who based it on the dissection of animals. It was only in the 1500s when Andreas Vesalius started dissecting human corpses that we started to understand our own anatomy.

When it came to seeing the ribosome, however, none of the methods we were using would let us visualize details of the ribosome, let alone how it worked. Before returning to our story, it is worth making a detour into how scientists took half a century to develop a technique that would be key to cracking the ribosome problem.

For most of human history, we were limited by what we could see with the naked eye. The detail we could see dramatically expanded in the mid-1600s when a Dutch linen merchant, Antonie van Leeuwenhoek, wanted to examine cloth fibres more closely. In his desire to make better lenses, he would go on to invent the most powerful microscope of his time, which he used to look at everything from pond water to the scum scraped off his own teeth.

He was astonished to see small creatures moving around, which he called animalcules but which we know today as microbes. Soon afterwards, Robert Hooke also used microscopes to look at the details of everything from fleas to various tissues. He coined the term *cell* to describe the compartments that made up plant tissues. The idea of the cell completely transformed biology. We now realize that the cell is the smallest unit of life that can exist independently, and cells can associate to form tissues and whole animals. As microscopes became more powerful, people saw that the insides of cells had structures, such as a nucleus with chromosomes and various organelles. Just being able to see detail had transformed biology from the anatomy of humans to the structures inside a cell. But what were all these things inside cells made of?

Like all everyday matter, cells and their components are made of molecules, which are groups of atoms that are held together in a very specific way. The atomic theory of matter took such a long time to develop and is so important that Richard Feynman, the famous physicist, said that if all scientific knowledge were to be destroyed and only one sentence could be passed on to the next generations of humans, it should be 'all things are made of atoms – little particles that move around in perpetual motion, attracting each other when they are a little distance apart, but repelling upon being squeezed into each other.'

It is astonishing that in the eighteenth and nineteenth centuries, without ever being able to see molecules, scientists not only deduced their existence but also their structure – the arrangement of the atoms that make up the molecule. They could do this for simple molecules like common salt, which only has two atoms, and somewhat more complicated molecules like sugar, which has a couple of dozen. But with larger and more complex molecules, it becomes increasingly difficult to infer their structures without being able to visualize them directly.

The reason nobody had ever seen a molecule has to do with the properties of light itself. Light is made up of photons, which we know from quantum physics can have the properties of both particles and waves. The wave-like nature of light is the reason lenses and microscopes work. But this property also means that when light passes through a very narrow opening or around the edge of an object, it spreads out because of its wave-like nature in a process called diffraction. Normally, we don't notice this effect, but if two very small objects are close together, their images spread out and merge with each other, and someone looking through a microscope would see just one large fuzzy object rather than two distinct ones. In the nineteenth century, the German physicist Ernst Abbe calculated that you could only see two objects as separate or 'resolve' them if they were no closer than half the wavelength of the light used to look at them. Visible light typically has a wavelength of 500 nanometres (a nanometre is one-billionth of a metre). Any features that were closer together than about 250 nanometres wouldn't be seen as separate. This minimum distance between two objects that could be seen as distinct, or resolved, is called the resolution limit. When using light, very fine details, such as features that are closer than 250 nanometres, simply wouldn't be seen but would be blurred out.

People had estimated by the early twentieth century how many molecules there would be in a volume of material, so they knew the approximate distance between the atoms in a molecule. It turned out to be over a thousand times smaller than the wavelength of light. This meant that it would be impossible to see individual molecules even with the best light microscopes. Molecules would be forever invisible.

An alternative to light emerged when a curious new radiation was discovered in 1895 by Wilhelm Röntgen, a German physicist who was looking at discharges from vacuum tubes. These tubes have two electrodes separated by a high voltage in a vacuum. When a current is applied, the negatively charged electrode or cathode heats up and emits electrons, which then fly through the vacuum and hit the other electrode, the anode. He discovered that these tubes emitted mysterious rays that caused a barium compound to glow even in total darkness. He called these mysterious rays X-rays

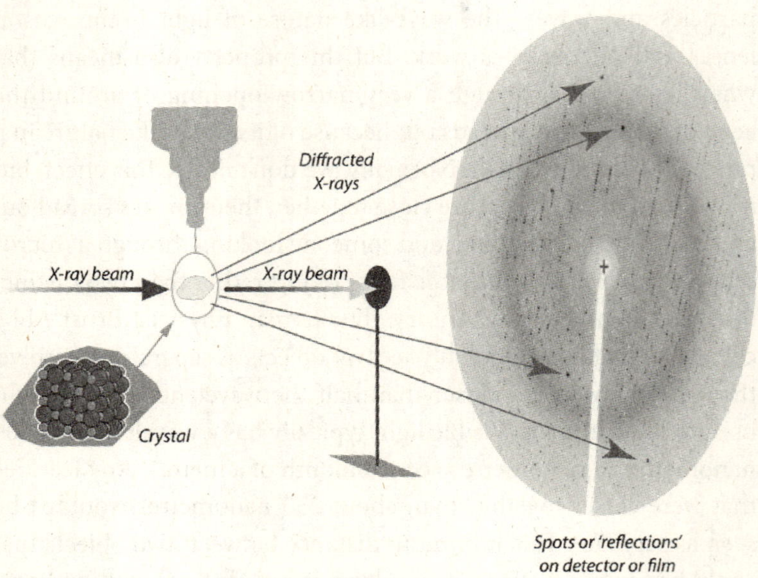

Figure 3.1 X-rays hitting a crystal to produce diffraction spots

and started to investigate their properties. They were highly penetrating and allowed people for the first time to see through normally opaque objects like the human hand, revealing the bones inside.

Nobody really knew what X-rays were or even whether they were particles or waves (we now know that they, too, are photons like ordinary light, so they are both particles and waves). In 1912, Max von Laue and two of his associates decided to see what would happen when X-rays hit a crystal of zinc sulphide, which consists of two types of atoms, zinc and sulphur. They found that the X-rays, instead of being scattered all over, were concentrated in spots.

Von Laue quickly realized what was happening. He had used a crystal, which he realized was just a regular three-dimensional arrangement of molecules, like a stack of perfectly spherical balls. When X-rays hit the crystal, if they were waves, each atom would scatter the waves in all directions, just as when you throw a pebble in a pond waves ripple outward in every direction. The resulting wave in any direction would be the sum of all the waves scattered by each atom that had been hit by the X-ray beam.

When two waves combine, the strength of the combined wave depends on how the two original waves add up, which in turn depends on how they are related to each other. If they each have their peaks and valleys in the same place, they are said to be in phase, and the combined wave will be twice as strong. But instead, if the peaks of one line up with the valleys of the other, they are out of phase and will completely cancel each other out. Anywhere in between will produce an intermediate result.

Von Laue realized that, depending on where each atom was located, the waves scattering off them would travel different distances. They would lag behind or be ahead, so they would no longer be in phase and would largely cancel each other out. But in certain directions, the waves from the different atoms would lag or be ahead by a whole number of wavelengths. In that case, their peaks and valleys would still line up, so they would still be in phase and reinforce each other. That's why von Laue saw just spots in his photograph – they indicated the directions in which the waves scattering off the atoms in the crystal had reinforced one another.

The experiment showed that X-rays could certainly be thought of as waves. But it was also the first direct proof that a crystal consisted of regular arrangements of atoms. Since people knew roughly how far apart atoms would have to be, this also told them what the wavelength of the X-rays was. It was just right to see atomic detail – more than a thousand times smaller than the wavelength of light. Two years later, in 1914, von Laue was awarded the Nobel Prize in Physics.

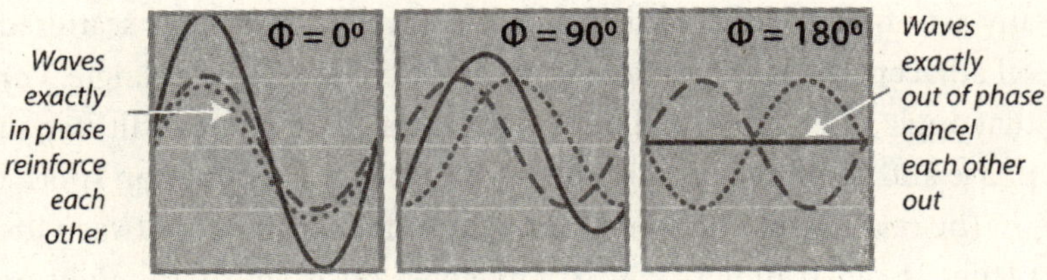

Figure 3.2 Adding waves depends on their relationship

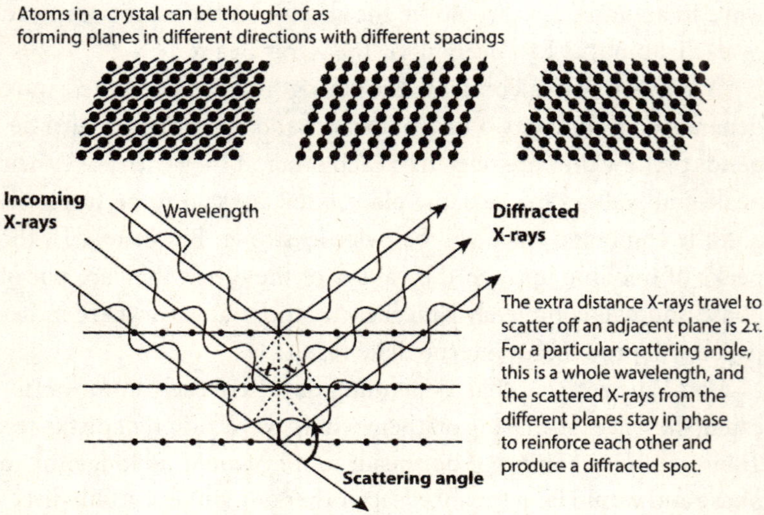

Figure 3.3 Planes in a crystal and how they diffract X-rays at certain angles

Von Laue had also tried to deduce exactly how the zinc and sulphur atoms in his crystal were arranged in space. But here, his analysis turned out to be wrong. In Cambridge, a young graduate student named Lawrence Bragg was intrigued by von Laue's results. He puzzled over them and came up with an elegant way of looking at the problem that helped deduce the correct structure. Bragg realized that the atoms in a crystal could be thought of as forming different sets of planes. These sets of planes could be in different directions and separated by different spacings. The X-rays scattering off the atoms in such a plane could be thought of as reflecting off the plane, so the diffraction spots are also called reflections. For any set of planes, the additional path travelled by X-rays scattered off adjacent planes is a whole wavelength for a particular angle. For that angle, the waves scattering from each set of planes will stay in phase and reinforce one another, giving rise to a diffraction spot.

The relationship between the angle and the distance between the planes is called Bragg's law. At any given position, there might be several planes that satisfy the Bragg condition, each giving rise to a spot at a particular angle relative to the incident X-ray beam. Also, as you rotate the crystal, new planes will satisfy the Bragg condition and give rise to more spots. When you have completely rotated the crystal about the beam, you will have measured all the possible spots from the crystal.

Using his analysis, Bragg worked out the correct arrangement of the atoms in von Laue's crystal. He reported his analysis to the Cambridge Philosophical Society in November of 1912, but because he was only a student, his professor, J. J. Thomson, who had discovered the electron, had to officially communicate the article Bragg wrote for the society's journal.

Bragg then used his theory to analyse one of the simplest molecules around, common salt. By then, chemists had concluded that a molecule of salt consisted of a sodium atom and a chlorine atom held together, which they called sodium chloride. When Bragg analysed the spots from his X-ray photographs of salt crystals, he found that there was no sodium chloride molecule. Instead, the crystal was a three-dimensional chessboard arrangement of sodium and chlorine ions (in which the sodium atom has lost an electron and the chlorine atom has gained one, so the two have opposite charges). The ions are then held in the crystal by electrical forces.

Many chemists of his time didn't take kindly to a young physics graduate student telling them that even a simple thing like common salt wasn't what they had thought it was. One of them, Henry Armstrong, a chemistry professor at Imperial College, London, viciously attacked Bragg in a letter to the journal *Nature* titled 'Poor Common Salt,' saying that Bragg's structure of sodium chloride was 'more than repugnant to common sense.' In what was perhaps the ultimate insult to an Englishman, he added, 'It is absurd to the nth degree, not chemical cricket.' In the end, Bragg not only turned out to be right but went on to determine the structures of lots of simple molecules. For the first time, molecules could be 'seen.' This method of determining the three-dimensional arrangement of atoms in a molecule by getting them to form crystals and analysing the diffraction spots came to be called X-ray crystallography.

Bragg's father, William Bragg (actually they were both William, so the son used his middle name, Lawrence), was a professor of physics and had developed some of the most advanced instruments of his time to make accurate measurements of the X-ray spots. Once Bragg had worked out the theory, he and his father worked together on several experiments. While Bragg remained in Cambridge, his father, who was already a famous physicist, was travelling widely giving lectures on his work with 'his boy.' For some time, Bragg was worried that since he was a mere student, his well-known father would get all the credit for their work, and there was apparently some tension between them. As it turned out, someone on the Nobel committee was very well informed. In 1915, both Braggs shared the Nobel Prize in Physics for their work. Bragg, who was then twenty-five, remains the youngest science Nobel laureate. He could not go to Stockholm to receive the prize because World War I had just started. In fact, Bragg's brother Robert had been killed in action only a few weeks before he heard of the award. Bragg ended up delivering his Nobel Lecture in 1922.

The simple molecules that Bragg initially studied had only a few atoms. So it was possible to try out different guesses for their structures and see if the spots predicted by Bragg's law would match up with what was actually seen in the photographs. But this sort of guesswork became more and more difficult for larger molecules that had many more atoms. At this point, a different approach was needed. Could you directly calculate an image or 'map' of the molecule from the X-ray data that would show you directly where the atoms are?

To understand how calculating a map works, think of how a magnified image is obtained using a lens. Light rays are scattered from each part of the object. Each point on the image is produced by having the lens combine the scattered waves from every point on the object. The important thing is that the scattered rays exist whether a lens is there or not; the lens simply collects them to form an image. We've discussed how the wavelength of light is almost a thousand times too large to see atoms in a molecule. X-rays, on the

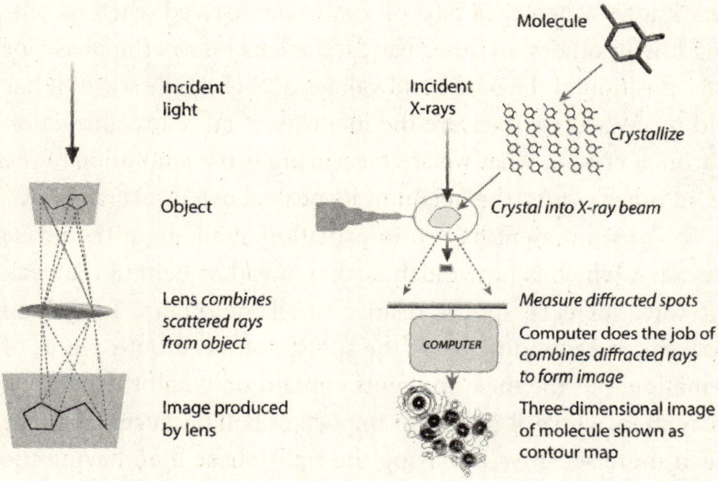

Figure 3.4 Comparing image formation by a lens and in X-ray crystallography

other hand, have the right wavelength. Couldn't we just use X-rays with a lens to see images of molecules directly and not bother with crystals and spots?

The problem is that there isn't a good enough lens to make images of molecules with X-rays. But even if we could do that, there is a serious problem because, unlike light, X-rays damage the molecules they hit. To see an individual molecule in sufficient detail, you would have to expose it to such a high dose of X-rays that you would end up destroying the molecule. In a crystal, however, the diffraction spots are the result of adding up the scattered X-rays from millions of molecules. The amplified signal from these millions of molecules means that you can get away with a much smaller dose of X-rays and is another important reason to use crystals.

Without a lens for X-rays, people figured out clever ways to do mathematically what a lens does – combine waves from different parts of the object into an image (for the mathematically inclined, this is doing a Fourier transform of the scattered rays). But there was a big problem with just taking the measured spots in an X-ray photograph and combining them in a computer to form an image.

A lens 'knows' when each part of a wave has arrived when recombining it with others. In other words, the lens knows the phase, or relative position of the peaks and valleys of each of the waves it has to add up. When we measure the intensity of an X-ray diffraction spot from a crystal, what we are measuring is the amplitude of the wave, in other words, the height of its peak above its average position. The measurement has no information at all about the phase of the wave, which is how much farther ahead or behind the peak of the wave for each spot is relative to all the others. To add up the waves corresponding to all the spots, you need both pieces of information, but the measurements contain only half the data. To make matters worse, it is the less important half because the image is much more sensitive to having the right phase than having the right amplitude. This vexing situation is called the phase problem in crystallography. Without knowing the phases, there would be no image of the structure.

The crystallographer Arthur Lindo Patterson came up with one way around the problem when he realized that, even without phases, you could use the measured intensities of the spots to calculate a function that would allow you to locate the most prominent of the atoms in the structure, which typically were the heavier atoms (because they have more electrons and scatter more). You could then calculate the phases that just these atoms would give rise to and combine them with the actual measured amplitudes from the whole molecule. When you do this, some of the missing atoms – the ones that were not part of your initial few atoms – show up as weaker features or 'ghosts' in the image of the structure. When you add those atoms to your initial structure and redo the calculation, then even more atoms show up as ghosts in the next round. In this way, you can gradually bootstrap your way to the final complete structure.

The end result of this is a three-dimensional image or map of the molecule. These maps are called electron density maps because almost all the scattering of X-rays is from electrons in the atoms, so the maps show how dense the electrons are at any given location.

In practice, since the electrons mostly form a tight core around the nucleus, the method tells you where the atoms are. The maps are visualized by making contour maps of sections, similar to topographical maps that show where mountain peaks are. In topographical maps, the higher the altitude, the higher the contour level; in electron density maps, the greater the density, the higher the contour level. So the maps show where the atoms in a molecule are.

Scientists started using the Patterson method to determine the structure of increasingly complicated molecules. One person who pushed this method to its limit was Dorothy Hodgkin (née Crowfoot). She was one of the first women to get a first-class honours degree in chemistry from Somerville College at Oxford University, and then went to get her PhD working under John Desmond Bernal in Cambridge.

Bernal was a brilliant polymath, but he was something of an intellectual butterfly. He often made the first pioneering forays into lots of important problems but didn't always see them to completion. Perhaps he had too many distractions. During World War II, he was asked by the British government to advise them on the best sites for the D-Day landings in Normandy. He was an ardent communist and remained an apologist for the Soviet government even after Stalin's atrocities became known. He was also ardent about women, on occasion having relationships with several of them at once. Many of them – including Hodgkin herself – sensed that Bernal genuinely cared about them and encouraged them in their careers, and remained on very good terms with him for the rest of their lives long after they were no longer involved with him. In fact, several of them took turns taking care of him when he was terminally ill.

Perhaps as a result of his various distractions, several of Bernal's protégés went on to make much greater contributions and become more famous. Hodgkin was one of the most illustrious of these. She returned to Oxford after her PhD, but the academic world was not at all receptive to hiring women, and she was unable to get a proper faculty position there. Luckily, her old college, Somerville, gave her a fellowship, which she supplemented with various temporary research grants. She was given space in the attic of the university's natural history museum, and to carry out her experiments there, she frequently had to precariously balance her precious crystals in one hand while she climbed up a ladder. Undeterred by her difficult and uncertain working conditions, she showed exceptional judgement in choosing to work on important biological molecules, including penicillin and vitamin B12. The latter contained several hundred atoms, and the determination of its structure was considered a tour de force. At one point, Bernal told her she would win a Nobel Prize. She asked whether she might one day be elected a fellow of the Royal Society, and he is reported to have said, 'That would be much more difficult!' For men, it would have been the other way around, but at the time, the Royal Society had not elected any women in its nearly three hundred years of existence. But Hodgkin's work was simply far too important to ignore. She was made a fellow of the Royal Society in 1947, just two years after the Society welcomed its first female fellows, the crystallographer Kathleen Lonsdale and the biochemist Marjorie Stephenson. In 1964, Hodgkin went on to win the Nobel Prize for her work, an event reported with the headline 'Nobel prize for a wife from Oxford,' beginning the article with: 'A housewife and mother of three yesterday won the Nobel Prize for chemistry.' Clearly, for at least some journalists, her domestic status and ability to procreate were still the most important facts about her.

X-ray crystallography had been a resounding success, but it wasn't clear initially that molecules like proteins could be studied by crystallography at all. In the mid-1930s, when Bernal and Hodgkin first examined crystals of a protein with X-rays, they hardly saw any spots. Bernal realized that protein crystals contain quite a lot of water and were drying out and losing the regular arrangement they originally had had. When he and Hodgkin kept them hydrated while doing their experiments, they saw a beautiful diffraction pattern. It was the first proof that proteins could have a defined structure and weren't random chains of amino acids.

But proteins contain thousands rather than hundreds of atoms, so the methods that Hodgkin had used to solve vitamin B12 would not work. Fortunately, Max Perutz, an immigrant from Austria, decided to take on this daunting problem. Perutz had left Austria in the 1930s, escaping the Nazis by only a few years. Like Hodgkin, Perutz went to Cambridge to work with Bernal, who by that time was known as Sage because he seemed to know everything. Perutz arrived in Bernal's lab just after Hodgkin had left and began working on haemoglobin, a large protein in our blood that consists of four separate chains, each with an iron atom that carries oxygen from the lungs to our tissues. It was around fifty times larger than any molecule that had been solved by crystallography at the time, and people dismissed Perutz as crazy. Perutz himself had no idea how he was going to solve it. He would proudly show beautiful diffraction pictures from his crystals to his colleagues, but when they asked him what it meant, he would quickly change the subject. Luckily for Perutz, Bragg, who became the Cavendish professor in Cambridge in 1938 and thus had great influence, was very enthusiastic about his goals and supported him for years, even though progress was slow or non-existent.

Eventually, almost twenty years later in 1953, Perutz made his breakthrough. When he added a heavy atom like mercury to his crystals, it changed the intensity of the spots. The heavy atoms bound to only a few places in the molecule, and by measuring the differences they caused in the intensities of the X-ray diffraction spots, it was possible to figure out where the heavy atoms were. They did this by carrying out the same Patterson calculations that Hodgkin had done, but this time on the difference in intensities between the crystal with and without the heavy atoms bound. The calculations would reveal where these heavy atoms were, which in turn would allow them to determine the phase for each spot and calculate a three-dimensional image of the molecule. Using exactly this method, over the next six years, Perutz and his former student John Kendrew solved the structures of haemoglobin and a smaller related protein that also carries oxygen, called myoglobin.

So around 1960, nearly fifty years after X-ray crystallography had revealed the structure of common salt with its chessboard arrangement of just two types of atoms, the technique could show what a protein with thousands of atoms looked like in three dimensions. The age of structural biology had begun.

Perutz was Crick's PhD supervisor, and Kendrew, at least officially, was Watson's postdoctoral adviser. Perhaps not entirely by coincidence, Perutz and Kendrew shared the Nobel Prize in Chemistry in 1962 for the first protein structures the same year that Watson and Crick shared the Nobel Prize in Physiology and Medicine with Maurice Wilkins for their work on DNA. That same year, Perutz moved his unit from the refurbished bicycle shed behind the Cavendish lab in the middle of town, where they had been tolerated by the 'real' physicists for many years, to their new home in a new four-storey building in the southern outskirts of Cambridge: the MRC Laboratory of Molecular Biology, or LMB. With four Nobel Prizes in its first year, the LMB kicked off with a big bang.

CHAPTER 4

The First Crystals of the Machine

AS A RESULT OF THE heroic efforts by Max Perutz and John Kendrew, people could see for the first time how the thousands of atoms in a protein came together to form precise structures. In their case, they could even see the iron atom that bound the oxygen molecule in myoglobin and haemoglobin.

Crystals are basically ordered three-dimensional stacks of identical molecules. At one extreme, if a molecule has just one atom, making a crystal of it would be like making a regular stack of spheres like billiard balls – fairly easy. But a large irregular molecule with thousands of atoms will not stack up so easily because all the molecules would have to line up in exactly the same orientation. Any slight irregularity and you would disrupt the pattern. It would be like piling up toy railway engines into a perfectly aligned and regular stack. The problem is even harder because large molecules like proteins are not entirely rigid. They tend to be floppy with little loops and extensions that wave around. So it is amazing that proteins crystallize at all, and in general, the larger the protein, the harder it is to crystallize. Even today, nobody can predict exactly how a protein will crystallize or even whether it will do so at all. Considering how uncertain the process is, it was not at all clear something like the ribosome, which had hundreds of thousands of atoms, not just thousands, would even form crystals.

For molecules to form a well-ordered crystal, they need to be nearly identical so they can all pack the same way in a three-dimensional stack. People initially didn't know whether all the ribosomes from a particular source like a bacterium or a tissue of an animal had the same structure or even the same set of proteins. If they didn't, it was highly unlikely that they would ever form crystals. The first hint that ribosomes might have a defined structure came only a decade or so after ribosomes were discovered, when Breck Byers at Harvard was looking at what would happen to cells in chick embryos when they were cooled. He wasn't initially looking for ribosomes at all but was studying long filaments in cells called microtubules that are involved in lots of processes like cell division. While doing this in 1966, he noticed that ribosomes in these cooled chick embryo cells would come together in regular sheets. These sheets were only one ribosome thick, forming two-dimensional crystals rather than the usual three-dimensional ones. Max Perutz invited Byers to the LMB to work on his two-dimensional crystals. He visited twice in the 1960s and 1970s but apparently nothing came of it.

In the meantime, two young scientists at the LMB, Nigel Unwin and Richard Henderson, were figuring out a different way of determining the structure of biological molecules. Unwin was tall and thin with bangs that made him look like a Beatle, while the shorter Henderson looked boyish enough to pass as a teenager (exacerbated by his tendency to wear shorts and sandals). Both were energetic and eager to make their mark on science. Unwin and Henderson were working out how to determine the structure of bacteriorhodopsin, a protein that sits in the membrane of a salt-loving bacterium and generates energy from light. At the time, there was no good way to get membrane proteins to form three-dimensional crystals. These proteins sit in the oily environment of the lipid membrane that envelopes all cells and are therefore not soluble in water, so the traditional methods of crystallizing a protein would not work. Unwin and Henderson decided to work with two-dimensional crystals of the sort that Byers had seen and use the electron microscope to obtain their structure.

Like X-rays, electrons have a wave-like character and have an even shorter wavelength. They had been used to obtain atomic structures of substances like minerals and metals, but biological molecules have very low contrast – meaning that in terms of their scattering properties, they do not stand out much compared to the surrounding water or lipid membranes. To see them with sufficient detail, you needed to expose them to so many electrons that the molecules would be damaged and disintegrate before you could produce their structure. However, with two-dimensional crystals, Unwin and Henderson worked out methods of obtaining a structure by doing crystallography using an electron microscope with low doses of electrons.

In 1972, while they were just starting to work out their methods, Unwin came across a paper reporting that ribosomes formed regular two-dimensional arrays similar to what Byers had observed, only this time from the oocytes (the cells from which eggs develop) of a kind of lizard. He wrote to the author, Carlos Taddei, enquiring about these crystals but never received a reply even after several attempts. With what could only be described as grim determination, Unwin took a train all the way from Cambridge to Naples, found Taddei's lab and knocked on his office door. Eventually, Taddei came to the LMB for a while to work with Unwin. Apart from his curious reluctance to respond to Unwin's initial enquiries, he was eccentrically insouciant in other ways. He became well known in the LMB for blithely puffing away at his pipe in the lab and regularly setting off the fire alarms.

Unwin pursued the problem for a few years, and although he did get some information, it became clear that these two-dimensional crystals of ribosomes from lizard oocytes were not good enough to obtain a detailed atomic structure. So Unwin eventually gave up on the problem and moved on to other things. Both he and Henderson went on to do pioneering work on the structure of membrane proteins. Unwin's lizards, which were kept in the basement of the building, escaped and multiplied and could occasionally be seen wandering outside the building for years afterwards.

Even though the two-dimensional crystals of ribosomes from chick embryos and lizard oocytes proved to be a dead end, they were important because the very fact that ribosomes could crystallize at all, even if only in two dimensions, suggested that at least they had a defined structure. Could the ribosome ever form three-dimensional crystals of the sort that had been used to solve the structures of proteins like haemoglobin? By the mid-1970s, many protein molecules much larger than haemoglobin had been crystallized, including large protein assemblies and entire viruses. So even though ribosomal subunits were ten times larger than the largest molecule that had been crystallized to date, it was not so unreasonable to try to coax them into forming the kinds of crystals from which one day a structure could emerge.

One person who thought so was Heinz-Günter Wittmann, who with his wife, Brigitte Wittmann-Liebold, had worked on the genetic code using tobacco mosaic virus, which uses a single RNA molecule rather than DNA to store its genes. In 1966, Wittmann was made a director at the new Max Planck Institute for Molecular Genetics in Berlin. This meant he had enormous resources and oversaw a department called by its German name Abteilung. The papers from the department would have 'Abteilung Wittmann' as part of the address (nowadays Max Planck directors rarely confer their own name on their department, instead preferring to call it after the area of their research).

Once the Max Planck organization hired you as a director, it was virtually unheard of for them to fire you. This meant that Wittmann could try things that might take a very long time to work. In an almost stereotypically German way, Wittmann systematically organized his department to look at every aspect of ribosomes. Some of it was important at the time but mind-numbingly tedious, like purifying ribosomal proteins from lots of different species and laboriously sequencing each of them. DNA sequencing, developed by Frederick Sanger in 1977, superseded much of this work because it was much faster to sequence the gene for a protein than to sequence the protein itself. But Wittmann was also smart enough to know that structure was the key to understanding how the ribosome worked.

A few years after Wittmann started his department, a curious character appeared in the world of crystallography. He was a German named Hasko Paradies, and although a paediatrician by training, he had taken to working on crystallizing important molecules. It seemed as though there was nothing he had not crystallized. He had crystallized tRNA before anyone else, as well as many large protein complexes. The only problem was that his work did not stand up to scrutiny. When David Blow, a pioneer in crystallography of enzymes, saw an X-ray diffraction picture of one of the so-called tRNA crystals during a talk Paradies gave when he was working at King's College in London, he immediately recognized it to be chymotrypsin, a protein he had solved many years earlier. He confronted Paradies and in fairly short order Paradies left King's.

On the strength of his having 'crystallized' so many different important proteins, Paradies was offered a job in Wittmann's department, even though Wittmann was apparently told of the circumstances surrounding Paradies's departure from London. He stayed in Wittmann's institute until 1974, the year he published a paper on crystallizing ribosomes, and went on to a professorship at the Free University in Berlin. Some years later, in 1983, Wayne Hendrickson and several other leading crystallographers wrote a letter to *Nature* laying out the reasons for their conviction that key pieces of Paradies's research represented 'deliberate misrepresentation' and 'should be discounted.' Although he responded to the letter defending his work, Paradies left his post in Berlin soon afterwards for what he maintained were unrelated reasons. At this point, he disappeared from the world of biological structures.

Often, just being told something is possible breaks down a huge psychological barrier and spurs people to try things. So even though Paradies's claim of tRNA crystals was widely rejected, Brian Clark, who was working on the problem at the time, said he was spurred on by Paradies's claims and even found some of his suggestions useful in producing actual crystals of tRNA. Similarly, Wittmann, unaware (or unconvinced) of Paradies's failings and seeing his initial 'result' as promising, continued to be interested in having someone try to crystallize the ribosome.

One person who was game was Bob Fletterick, a crystallographer who was at the University of Alberta in Edmonton, Canada. He was about to get tenure there, but his girlfriend at the time was a native-born German, and he thought it would be good for them to spend a few years in Germany. So in early 1978, he contacted Wittmann to discuss the possibility of working on crystallizing the ribosome. Wittmann agreed it would be worthwhile and sponsored Fletterick for a Humboldt Fellowship, which was granted some months later. Fletterick recalled the salary as being higher even than for the faculty position he had in Canada. His stint with Wittmann never materialized, however, because Fletterick's girlfriend 'suddenly took a diversion,' so he no longer felt like going to Germany. By that time, he was getting lots of faculty offers in the US and ended up spending the rest of his career in San Francisco as a tenured faculty member at UCSF.

Having struck out with both Paradies and Fletterick, Wittmann might have been feeling discouraged, but he was lucky the third time around when a scientist from Israel, Ada Yonath, enquired if she could work with him. She had exactly the right combination of ambition and tenacity needed for this project.

At the time, Yonath was a faculty member at the Weizmann Institute in Israel. At some point after meeting Wittmann at a conference, she broached the idea of spending some time at his institute in Berlin. Fortunately, Wittmann still had the unused Humboldt Fellowship from his application with Fletterick, and he quickly arranged for it to be given to Yonath instead.

Yonath's road to Berlin was not an easy one, and she had to overcome some daunting barriers along the way. She grew up in

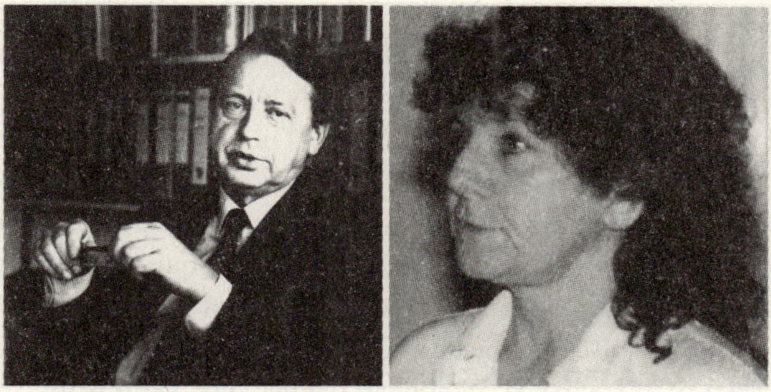

Figure 4.1 **Heinz-Günter Wittmann** *(courtesy of Brigitte Wittmann-Liebold)* **and Ada Yonath** *(courtesy of William Duax)*

Jerusalem, in a poor orthodox family whose circumstances were made more difficult by the premature death of her father at the age of forty-two. Yonath's parents encouraged her education, although her difficult circumstances meant that she had to work and help support the family from quite a young age. In an early glimpse of her determination, she managed to find the resources to go to university, went on to graduate from the Hebrew University of Jerusalem, and obtain a PhD from the Weizmann Institute. After doing a postdoctoral stint in the US, she returned to the Weizmann Institute to become a faculty member there.

Yonath began her ribosome-related work by trying to crystallize one of the protein factors that helps the ribosome start at the right point on the mRNA, but after a year, she had had little success with it. She suffered a setback after a bicycling accident and had to spend several months recuperating. During this time, according to an interview with Elizabeth Pennisi in *Science* in 1999, she realized that Wittmann's lab was making lots of purified ribosomes from different species and asked Wittmann if she could try to crystallize them.

Yonath's experience in crystallography was limited to a couple of small proteins, and she had no previous publications related to ribosomes. But after his two false starts with Paradies and Fletterick,

Wittmann was probably only too delighted that someone was willing to take on the extremely challenging project. In the same Pennisi article, Yonath recalled, 'He said this was the dream of his life and gave me everything I needed.'

Advances in biology are often made by choosing the right organism to work on. For example, the study of nerve transmission was made possible by using giant axons from squid because they were large enough to see and stick electrodes into. Early geneticists used fruit flies because they breed quickly and you can follow lots of visual markers like eye colour to deduce how various traits are inherited. For bacteria, the standard organism for biochemical and genetic studies of all sorts is *Escherichia coli,* or *E. coli* for short, because it is easy to grow and doubles every twenty minutes, and one could do genetics with it. Its name refers both to the person who discovered it, Theodor Escherich, and to its being found in the human colon. It is more familiar to the general public because of the occasional outbreaks of severe dysentery some virulent strains of it can cause. Not surprisingly, it became the main source for purifying and studying ribosomes, and Wittmann's lab had an abundant supply of the microorganism. The initial efforts to obtain crystals from *E. coli* ribosomes resulted only in small microcrystals – so small that they were no more useful than the two-dimensional crystals others had already been working on with electron microscopy. They needed a new organism, and luckily, their colleague Volker Erdmann had just the right one to get them started.

At age fifteen, Erdmann immigrated from Germany to the US where he went to high school and college in New Hampshire. He was curious to return to his roots, so he went back to Germany for his PhD. After his PhD, he went to work in Masayasu Nomura's lab in Wisconsin. He had heard of Nomura's work on taking apart and putting back together the entire small or 30S subunit of the ribosome and wanted to see if he could do the same with the large or 50S subunit. The names 30S and 50S for the subunits of bacterial ribosomes indicate how fast they sediment in a test tube when spun at high speeds in a centrifuge. The capital letter *S* stands for Svedberg units, after the Swedish scientist Theodor Svedberg, who characterized how molecules sediment in an ultracentrifuge. Oddly, the whole bacterial ribosome is 70S not 80S because how fast a particle sediments depends on its overall shape as well as its mass.

Erdmann initially tried to reassemble 50S subunits from *E. coli*, which is naturally what Nomura had used for the 30S subunit. But he failed completely. So he switched to a bacterium called *Bacillus stearothermophilus*. The name *stearothermophilus* means 'heat loving,' and the bacteria grow naturally in hot springs at around 60 degrees C (or 140 degrees F). When he took 50S subunits from these thermophilic bacteria apart, he was able to repeat the trick Nomura had done with the small subunit. After his fellowship, in the last of his oscillations between Germany and the US, he moved to Wittmann's department in Berlin where he set up his own lab and took his samples of 50S subunits with him.

One day in the late 1970s, Yonath and Wittmann told Erdmann about their plan to crystallize ribosomes and asked him if he would like to help. Erdmann said he could help them if they worked with *B. stearothermophilus* ribosomes, since he was familiar with them. Because molecules from thermophilic bacteria are resistant to heat, they are thought to be more stable and likely to crystallize. They decided to meet the following Sunday morning. Erdmann asked his wife, Hannelore, who worked with him, to come along because she knew where the old samples of the large subunit were in the freezer. Yonath set up crystallization trials on them with Erdmann and his wife. Just three days later, on Wednesday, Erdmann recalled, Wittmann told him they had crystals. Barendt Tesche, an electron microscopist, confirmed that they were indeed crystals of the 50S subunit.

The team tried to improve on the original small crystals. When Erdmann's original material ran out, they had to make large subunits from freshly made ribosomes. There was a period when they couldn't reproduce the original crystals. Erdmann told me jokingly that he worried that since the original crystals were from subunits that had been stored in a freezer at minus 80 degrees C for four years, they would have to wait a few years each time after they purified ribosomes to obtain crystals! In the end, however, they managed to get crystals routinely.

Getting real three-dimensional crystals of a molecule with several hundred thousand atoms was a major achievement. But instead of reporting it with great fanfare in one of the major journals, the results were published in a then new and now defunct journal called *Biochemistry International.* Wittmann was a founding editor of the journal, whose name belied the fact that hardly anyone read it. I asked Erdmann why Wittmann would choose a relatively obscure forum to publish such an important result instead of a high-profile journal like *Nature* or *Science.* He speculated that Wittmann was fairly conservative and also a little hesitant and cautious because of the Paradies episode, so he didn't want to make a big deal of it but wanted to at least put it out on record.

Buoyed by their initial success, Wittmann did his best to ensure long-term support for Yonath to continue working on the problem. He tried to get her a director's position like his own, but the august Max Planck Society, apparently unimpressed by Yonath's credentials at the time, declined. However, he did eventually persuade the Society to support a special lab for her in Hamburg, right next to the German synchrotron, whose powerful X-ray beams would be needed to study these crystals. Just as importantly, he provided extensive support from his own department for making and characterizing the ribosomes. Over the years, he and Yonath became close friends.

Their success motivated others to enter the fray. The small town of Pushchino was built by the Soviet government as a centre of science. It was home to several well-funded institutes, one of which was headed by a brilliant ribosome biochemist named Alex Spirin. Like Wittmann, Spirin too was directing a large group studying virtually every aspect of ribosomes. He was not as systematic as Wittmann but was a highly imaginative scientist who would publish bold new ideas because he was unafraid to be wrong occasionally. Spirin was also a highly independent man who didn't like to bow to authority. One example of this was when he was asked to sign a petition to expel Andrei Sakharov, the dissident nuclear physicist and father of the Soviet hydrogen bomb, from the Soviet Academy of Sciences. Openly refusing to sign the petition might have been politically awkward for a prominent member of the Academy who was also the director of a major institute, so Spirin decided to become unavailable by going on a long 'hunting trip' in some remote forests outside Pushchino.

Spirin's institute was also interested in the structure of ribosomes, and one member, Maria Garber, ran a small group that was trying to crystallize individual proteins of the ribosome or the protein factors that helped the ribosome do various tasks. Like everyone else, she, too, tried proteins from *E. coli* ribosomes.

Garber changed the course of ribosome research in 1978 when she read a report from Japan that described using a new bacterium that grew at even higher temperatures than *B. stearothermophilus* as a source to crystallize two important proteins that acted on the ribosome. It was isolated from the hot springs of the Izu Peninsula in Japan in 1971 by Tairo Oshima and grew best at the searingly hot temperature of 75 degrees C (or 167 degrees F, which would only take a few seconds to burn your hand if you were so foolish as to plunge it into a hot spring at that temperature). They gave the new bacterium the unmistakable if tautological name of *Thermus thermophilus.*

Garber decided that this new bacterium was the answer. She went to Japan for a few months and brought back some stocks of the bacteria in December 1979, but unfortunately, they had died en route. She asked Oshima to send her some fresh cells by mail, which arrived safely. By the end of 1980, Garber and her colleagues had successfully used these bacteria to obtain beautiful crystals of a large protein, or factor, called elongation factor G, which helps the ribosome move along the mRNA.

This initial success with *T. thermophilus* proteins encouraged Garber and her colleagues to try other things with the organism. Given the resources in the Soviet Union at the time, it was quite expensive to grow large amounts of *T. thermophilus*, and Garber

Figure 4.2 Maria Garber with her group in Pushchino, Russia. Marat Yusupov is at the top right *(courtesy of Maria Garber)*

didn't want to waste any of it. They invited other scientists from the USSR to take whatever proteins they wanted from the bacteria, like a slaughterhouse using every part of an animal that is killed.

One of Garber's colleagues was Igor Serdyuk, a Communist Party member involved in 'communication with foreign countries' who had no problem travelling frequently to the West even at the height of the Cold War. He had previously used low-resolution techniques to characterize the overall shape of ribosomes, so he was naturally interested in seeing if he could crystallize them with material from Garber's laboratory. He and his student Liza Karpov purified ribosomes from *T. thermophilus* and obtained very small crystals, like the very first crystals in Berlin. With this initial success, Garber asked Spirin if he would support a group to try to crystallize ribosomes from this new organism.

He agreed, and she was joined by several people, of whom the most notable was Marat Yusupov, a student of Spirin's. Since they didn't have a lot of crystallization expertise, they asked two people from the Institute of Crystallography in Moscow to help them, Vladimir Barynin and Sergei Trakhanov. By 1986, they had crystals of both the small subunit and, using a trick Trakhanov had used to purify the ribosomes, the entire ribosome. Along with the 50S crystals from Yonath's and Wittmann's efforts, this meant that both subunits and the entire ribosome had now been crystallized.

Yusupov presented their results as a poster at a meeting in July 1987 in Bischenburg, near Strasbourg, France, and a month later the work was published in the European journal *FEBS Letters*. Within a few months, Yonath and Wittmann reported that they too had crystals of the small subunit and whole ribosomes from the same *T. thermophilus* species that the Russians had used. They published their results in the same obscure *Biochemistry International* journal, where they had reported their original crystals several years earlier. The following year, Yonath reported improved crystals of the small or 30S subunit that looked at least as good as the Russian ones.

This might have led to a head-to-head competition between the Russian and German groups, but it was not to be. In comparison with the Germans, the Russians were poorly funded and equipped, especially for crystallography of large molecules. In an effort to take the work to the next stage, Marat Yusupov and his wife, Gulnara, went to Strasbourg to continue the ribosome crystallography work in collaboration with Jean-Pierre Ebel and Dino Moras. For reasons unknown to Yusupov, at some point Ebel decided to discontinue the collaboration. For his part, Spirin believed that Ebel was discouraged by Yonath and Wittmann from competing with them.

Whatever the reason, the Russian effort on crystallization of whole ribosomes fizzled out, and Maria Garber went back to her original interest in individual ribosomal proteins and factors. With their efforts frustrated, some of the key people in the Russian effort dispersed all over the world. Some years later, in the mid-1990s, Yusupov wrote to Harry Noller, a leading ribosome biochemist at UC Santa Cruz in California, asking to work on the structure of the ribosome in his lab, but that is a story best told later. Sergei

Trakhanov seems to have led a peripatetic life, doing various jobs in Japan and the US for almost two decades. He too worked for some time in Noller's lab, after Yusupov had left it, before returning to Europe.

With the Russian effort essentially closed down, Yonath was leading the only group working on ribosome crystallography. By the end of the 1980s, none of the crystals was good enough to yield an atomic structure of either subunit of the ribosome, let alone of the entire object. But they could in principle have been good enough to reveal something about how the proteins and RNA were organized.

Instead, progress on the structure of the ribosome was slowly emerging from the fuzzy images of the ribosome seen by electron microscopy. Some of the work involved antibodies, which are proteins produced by our immune system that can bind to very specific targets. In one clever experiment, Jim Lake from UCLA, one of the scientists studying ribosomes using electron microscopy, made antibodies that recognized the beginning of a newly made protein. He showed in 1982 that these antibodies were attached to the back side of the large subunit. This was on the opposite side from where the new amino acid on the tRNA was attached to the growing protein chain. This suggested that there must be a tunnel in the large subunit: a birth canal through which a newly made protein chain would have to travel before it emerged on the other side. A few years later, in 1986, Nigel Unwin confirmed the presence of this tunnel by analysing his two-dimensional crystals of lizard ribosomes in the electron microscope. The following year, Yonath and Wittmann also reported that there was a tunnel by analysing two-dimensional sections of their crystals of ribosomes in the electron microscope. Both of these reports were of fuzzy or low-resolution structures. Also, neither the ribosome nor the tunnel in these structures bear a strong resemblance to the ribosome we know today. It is not at all obvious that either group would have been so confident in identifying features in their maps as the tunnel through which the newly made protein chain emerges, if Lake had not already shown it existed.

Apart from these limited results, progress was slow. Even a decade after the first three-dimensional crystals of the ribosome had been produced, it was not obvious that they could ever be used to produce *any* meaningful images by X-ray crystallography. Certainly, an atomic structure of the ribosome seemed like a pipe dream. Nevertheless, even when it was not at all clear that it would ever be technically feasible, Ada Yonath kept the dream alive by continuing to explore different species and conditions to improve her crystals.

CHAPTER 5

Going to the Mecca of Crystallography

MEANWHILE, IN THE MID-1980S, I was slowly becoming frustrated at Brookhaven trying to milk what I could using the techniques I knew. They would never produce the kind of detailed images that could lead to an understanding of how the ribosome worked or even how chromatin was organized. Luckily, a few years after I got to Brookhaven, I was joined by Steve White, who had come from Wittmann's department in Berlin. Steve had grown up in a working-class neighbourhood in East London, and the odds were stacked against his emerging out of it. But he was smart and lucky enough to study at a grammar school (a selective government-funded school in England that admits pupils on the basis of an entrance exam taken at the age of eleven).

From there, he went on to get an undergraduate degree from Bristol and a PhD from Oxford. He was a sociable and friendly man who shared my bawdy sense of humour and like many Englishmen enjoyed a beer with friends and liked sport of all kinds. When he arrived from Berlin, he was involved with a South Indian woman who had moved to Philadelphia, but the relationship that had endured a trans-Atlantic separation did not survive the much shorter distance to Long Island. With its abysmally low ratio of women, Brookhaven was not exactly promising for the dating scene if you were a single guy, but somehow Steve, with his English accent, outgoing manner, and charm, managed to enter into a series of relationships during his time there.

In Berlin, instead of working on entire ribosomal subunits like Ada Yonath, Steve and his fellow Englishman Keith Wilson were working with Kryztof Appelt and others on the much more feasible goal of solving the structures of individual ribosomal proteins. After a long hiatus, they had solved the second one, leaving only forty-eight more to go. But even if they had solved all of them, it would be like knowing what lots of the peripheral parts of a car looked like, such as a fuel line or a spark plug, without any idea of how they fit together in a working vehicle. Of course, the two thirds of the ribosome that was RNA would still remain completely unknown. But it was better than nothing. I suppose what we hoped was that the structures would provide some clues about how they fit into the ribosome and what they did. Steve was keen to carry the project forward after he arrived there.

Steve was a well-trained crystallographer who had already been involved in studying several ribosomal proteins. I, on the other hand, knew a lot about how to take a ribosome apart and purify its proteins but knew no crystallography at all. Given our shared interest in the ribosome, Steve suggested we work together. I was delighted and thus began a collaboration that lasted almost fifteen years and changed my life.

Initially, I went to Yale to use their giant fermenter to grow large amounts of the same *Bacillus stearothermophilus* bacteria from which the Berlin group had obtained their first 50S subunit crystals, and from which Steve and his colleagues had crystallized several ribosomal proteins. But after seeing how tedious the whole procedure was and how little pure protein we got from it, I felt there had to be a better way. As luck would have it, I was in exactly the right place at the right time. My colleagues Bill Studier and John Dunn were busy figuring out how to trick the standard *E. coli* bacteria into making large quantities of almost any protein they wanted. They did this by using a special signal used by a virus called T7 that attacks *E. coli* and enables it to hijack the bacterial apparatus to make its own proteins. They figured that they could get *E. coli* to make large quantities of any desired protein by introducing the special signal from T7 to the beginning of the gene for that protein. So I told Steve I was going on a little sabbatical in our own department to learn the tools of molecular biology, and he should just wait for a while.

My lab at the time consisted of two technicians, Sue Ellen Gerchman and Vito Graziano, who were joined a few years later by a third, Helen Kycia. Under Bill and John's tutelage, Sue Ellen and I quickly cloned the genes for all the proteins that Steve White and his Berlin colleagues had crystallized. Soon after that, we made lots of the proteins and were delighted when these genetically engineered or recombinant proteins made in *E. coli* crystallized exactly like the ones purified by direct extraction from ribosomes from the original thermophilic bacterium.

At this point, I was also working on a key protein in chromatin that helps to condense it into a filament and pack it in the nuclei of cells. The protein, called a linker histone, has a central core called GH5. I wanted to crystallize it but had no experience in how to obtain crystals. Steve said, 'It's very straightforward. Let me show you how.'

Any schoolchild knows that if you let a solution of salt or sugar dry out, you get crystals. This is because as it dries out the salt or sugar becomes so concentrated that it is no longer soluble, and the molecules pack together to form crystals that come out of solution. If you let proteins dry out, though, they just form an amorphous goop, not crystals, because they are large and floppy and there are just too many ways for them to pack against each other. This is what happens when you add lemon juice to milk and the proteins quickly curdle. To coax large protein molecules to come out of solution in the form of crystals, you have to increase their concentration very slowly, giving the molecules the opportunity and time to line up in an orderly way to crystallize. This was typically done by mixing a drop of a protein solution with a small amount of a precipitant, like alcohol or salt, which makes the protein insoluble. This drop is then placed on a thin glass slide called a cover slip and inverted over a little well containing a solution that has a lot more salt or alcohol or whatever it is that makes the proteins insoluble. Water molecules leave the drop as vapour and enter the solution in the well until the salt concentration is the same in both the well and the drop. As the drop slowly shrinks because water is leaving it, the precipitant and proteins become concentrated enough to make the proteins insoluble. If it is done slowly enough, and if the conditions are just right, the proteins will come out of solution by packing regularly against each other to form crystals. In those days, we had to do all of the manipulations by hand, but today, robots routinely set up thousands of trials by varying the composition of the protein solution to try to find a condition that produces crystals.

Steve helped me set up a few different ways of trying to get crystals, and we soon had crystals of GH5. In the meantime, we had also reproduced crystals of a ribosomal protein called S5 (the prefix and number meant it was roughly the fifth largest protein from the '(S)mall' subunit).

Having obtained crystals of these proteins, I didn't want to just sit back and watch Steve solve them. I wanted to learn how to solve them myself. But I didn't know any crystallography, nor did I know how hard it would be to learn it, let alone become expert in it. Steve was reassuring: 'With your background in physics, you'll find this stuff dead easy by comparison.' That gave me the courage to embark on yet another venture, but how to go about it?

For a start, I took a crash course in crystallography at Cold Spring Harbor Laboratory, a well-known lab just thirty miles west of Brookhaven that was headed by none other than Jim Watson. In addition to conducting research and organizing meetings, the lab ran short specialized courses for scientists to pick up new techniques, often taught by world-renowned experts. In 1988, they had just started a new two-week course in crystallography, and I figured it would be a great way to quickly acquire the basics of the method. I was taught by some of the greatest scientists in the field, and on a free day, I took one of them, Hans Deisenhofer, on a long walk to see Teddy Roosevelt's home a few miles to the north of the lab. Hans was wearing totally unsuitable dress shoes and developed painful blisters on the way back. He probably forgot about his pain two days later when he shared the Nobel Prize for determining the structure of the protein complex that captures energy from sunlight and converts it into chemical energy – one of the most fundamental reactions of life.

A year later, my department was considering whether to give me tenure. If they didn't, then I would be out of a job and learning crystallography would be the least of my worries. I had produced a few good papers, mostly using neutron scattering. But by that time, I had already concluded that the technique produced almost no information of real relevance to how molecules actually worked. My work had reached a dead end, and crystallography was clearly where the action was. Hardly a week went by without a new article on how the atomic structure of some important molecule had forever changed the understanding of that field. Moreover, the tentative initial forays I'd made with Steve's help and the course I'd taken had whetted my appetite.

A tenure committee grilled me on my long-term plans. I took a deep breath and told them that, if they gave me tenure, I'd stop doing what I'd been doing, which was using neutron scattering, and immediately go away on sabbatical for a year to learn crystallography. To my immense relief, they agreed it was a good idea. A few days later, the senior staff of the department met, and that evening, John Dunn came to my home and handed me a long stick covered with aluminium foil, saying, 'Welcome to the tenure staff!'

When I began thinking about where I should go on sabbatical, there was really only one place I considered: the LMB in Cambridge. It was the birthplace of protein crystallography, and of course Cambridge was the birthplace of all crystallography. I also knew lots of Americans had gone there on sabbatical and enjoyed it. On a personal level, Vera and I were Anglophiles who loved English literature and culture. We religiously watched *Masterpiece Theatre*, and I loved the eccentric off-the-wall humour of Monty Python.

The director of the LMB at the time was Aaron Klug, a towering figure in structural biology. Aaron had led one of the two groups that solved the structure of tRNA, the other being a collaboration led by Alex Rich at MIT and Sung-Hou Kim at Duke. As is often the case with high-stakes competition, that race to obtain the first tRNA structure ended up being quite acrimonious. Aaron was a protégé of Rosalind Franklin's and at first sight could be mistaken for Woody Allen. He was also a leading figure in chromatin structure, and I thought it would be great to work in his lab. So I summoned up my courage and wrote to Aaron, saying that I had crystallized the linker histone involved in compacting chromatin and would like to come to the LMB on sabbatical to learn crystallography so I could solve its structure. I was pessimistic that he'd be interested in a nobody like me but was delighted when he wrote back a few weeks later encouraging me, saying he would be happy to sponsor me for a Guggenheim fellowship. With the fellowship and a half-salary from Brookhaven, I was all set to go off to England for a year. I didn't want to embarrass myself with Aaron after he had taken a chance on me, so I thought I should go prepared with some real data that I could learn to analyse on my sabbatical.

Despite the crash course in Cold Spring Harbor, I really didn't know the details of collecting and processing data, but here another colleague came to the rescue. Bob Sweet had grown up in small-town Illinois before going to Caltech for his undergraduate work. After his PhD from Wisconsin, he went to the LMB to do a postdoc like so many Americans, before spending time as a faculty member at UCLA. Like many Midwesterners, Bob too was an Anglophile, and his stint in Cambridge had only made it worse – he would routinely use Anglicized vocabulary, grammar, and spelling. Some years earlier, he and I had arrived at Brookhaven the same week and were living in temporary housing on site. When I first saw him, I was instantly struck by his large, luxuriant, and meticulously tended Poirot-like moustache; as he became balder over the years, it became an even more prominent feature. He could be sarcastic with a pedantic, avuncular manner that put some people off, but I found him a thoughtful, warm, and generous person. We went on to become good friends.

Bob ran a crystallography instrument or beamline at the X-ray synchrotron at Brookhaven. He took me under his wing and taught me a huge amount about how to collect and process data from crystals. Bob also got me interested in a crystallographic technique that turned out to be crucial later for solving the ribosome. Ever since Max Perutz and John Kendrew had solved the first protein structures thirty years earlier, the main way to solve structures was to collect data on the protein crystals and then repeat the process by soaking different compounds of heavy atoms like gold or mercury into the crystals. The differences between the data with and without the heavy atoms would reveal their positions and allow you to determine the phases of the X-ray reflections. With these phases and the measured intensities, you could calculate the structure. Getting crystals of a protein was uncertain enough, but getting these heavy-atom derivatives introduced yet another uncertainty. Often, soaking heavy atoms would destroy the quality of the crystals. Or they wouldn't bind to the proteins at all. The heavy-atom method became known as 'soak and pray.' But just as I started doing crystallography in the late 1980s, a new method called multiwavelength anomalous diffraction or MAD began to produce some promising results.

The Dutch crystallographer Johannes Bijvoet had established the principles of MAD in 1949. It was based on the fact that some atoms can absorb and re-emit X-rays instead of scattering them right away. As a result of this anomalous scattering, there are small differences in the intensities of pairs of diffraction spots that should be completely identical because of the symmetry of the crystal. The pairs of symmetric spots are called Friedel pairs, and the difference in their intensities – the anomalous differences – contain phase information, just like the differences as a result of adding a heavy atom to a crystal. Normally, atoms in a biological molecule like carbon, nitrogen, or oxygen do not have much anomalous scattering, so any differences are typically too small to be useful. But in 1980, Wayne Hendrickson, then at the Naval Research Lab, had solved the structure of a small protein using just the anomalous scattering from the unusually large number of sulphur atoms in the structure (which came from the amino acid cysteine).

Around this time, synchrotrons started being used for X-ray crystallography. They are large particle accelerators that accelerate electrons to close to the speed of light. As the electrons orbit around, they emit extremely intense beams of X-rays that could be used for diffraction studies. Keith Hodgson and his colleagues at Stanford realized that with synchrotrons, you could also choose the wavelength of the X-rays incredibly precisely. This meant you could collect data at two different wavelengths, where the scattering of some special atoms would change significantly. The difference between the two data sets would give you the position of the special atoms, and you could then calculate phases just as you might with heavy atoms. Moreover, you could do this in a way where at one of the wavelengths, the anomalous scattering of the special atom was particularly large.

Wayne Hendrickson developed an elaborate formalism to do this that was different from the one Hodgson had outlined and went on to solve a series of structures with it. Wayne also came up with the brilliant idea of growing bacteria in which the sulphur atom in the amino acid methionine was replaced with selenium, so that all the proteins would have selenomethionine wherever there would normally be a methionine. Not only was the anomalous scattering of selenium much larger than that of sulphur at typical X-ray wavelengths, but it had a peak at a wavelength that was very convenient for synchrotron instruments – around 1Å (or 0.1 nanometres). This made the method incredibly powerful. In principle, any protein that had a sufficient number of methionines ought to

Figure 5.1 The author and Steve White watching the data pour in at the Brookhaven synchrotron *(courtesy of Robert M. Sweet)*

be solvable this way, and today it is one of the most common ways of solving a brand-new protein structure.

Bob was interested in getting the method to work on his instrument at the Brookhaven synchrotron. So Vito Graziano produced crystals of GH5 with selenomethionine, and with Bob's help, we carefully collected data around the wavelengths where the selenium atom changed its properties. Steve and I also collected data on S5, but by the traditional heavy-atom method where we had soaked a gold compound into the crystals. I now had complete data on two proteins but no idea what to do with them. It was time to head for England.

My family and I took off for England in late August 1991. When we landed, I rented a large people carrier and packed the four of us and all our luggage including three bicycles into it. Despite having to drive on the left side of the road after an overnight flight from New York, I managed to find my way to Cambridge without an accident. When we got to the Addenbrooke's Hospital site, I was lost in its labyrinthine maze of one-way streets and asked passers-by where the MRC Laboratory of Molecular Biology was. To my great surprise, the first few people I asked had no idea where the world-famous lab was! This immediately reminded me of the story in Crick's autobiography *What Mad Pursuit* in which his taxi driver had never heard of the Cavendish lab even though scientists the world over had known it for over a hundred years. It made me realize that fame among scientists has a very narrow reach.

Aaron had assigned his longtime colleague John Finch to be my official host. Like Aaron, John too had been in Rosalind Franklin's lab and had moved with Aaron to Cambridge when the new LMB building had opened. When I arrived, John told me that unfortunately they did not have a place for me to work just yet. I rather naïvely told him that all I needed was a small desk in a corner of his lab. John politely smiled at this, and I found out a day later that he, a world-famous scientist, only had a desk and a small part of a lab bench!

This was typical of the LMB at the time: many senior scientists didn't have their own lab and often just had a desk in a shared lab or office. The lab was incredibly crowded, with equipment in the hallways and almost no free space. Crowding people together may well have been responsible for the LMB's success because it made people talk to each other and share ideas and techniques.

On my first full day there, I arrived around 9 a.m., and about an hour and a half later, John came by and asked me if I would like to go up to the canteen for coffee. I thought I had hardly done anything yet and declined, saying I didn't drink coffee. Again, John gave me one of his enigmatic smiles, and a colleague who watched this exchange, said, 'He hasn't learned our ways yet.' As the days passed, I realized that these regular breaks from work to have meals or coffee or tea allowed scientists to talk informally in the famous canteen on the top floor – a feature that has since been replicated in lots of scientific buildings. Humans can only really concentrate hard for a couple of hours at a time, and these short breaks re-energized people. The canteen was especially wonderful for a sabbatical visitor because I quickly got to know lots of scientists and made several lifelong friends.

My year at the LMB made me realize what a special place it was and changed my entire outlook on science. Not surprisingly, many scientists all over the world view it as a model for how science should be carried out, even if they don't always persuade their own institutions to adopt its character; a notable exception is the Janelia Research Campus of the Howard Hughes Medical Institute, which explicitly modelled itself on the LMB and Bell Labs. I found that unlike the vast majority of scientists, almost nobody at the LMB was working on routine problems just because they would lead to publishable results. Rather, they were trying to ask the most interesting questions in their field and then developing ways to answer them. A simple but telling question they would ask each other was, 'Why are you doing this?' Another lesson was that even very famous scientists like Max Perutz or Aaron Klug would unabashedly ask questions at lectures that were often trivial to people in the field. It made me realize that I shouldn't be ashamed of my ignorance and that no question is too stupid to ask if you want to know the answer.

A third lesson was that a lot of the LMB's success had to do with limiting the size of teams to just a few people. This forced the group leaders to focus on the most interesting questions and also participate or at least stay in close touch with the actual work. Today, there is often a tendency for famous professors to build up huge groups of twenty to thirty people simply because they can. This may be great for the professor, but it is not often a great environment for the people being trained, because many of them are relegated to problems of lesser interest and have less frequent mentorship. Not surprisingly, many studies also show that these very large groups are actually less productive for the cost than smaller ones.

In my first meeting with Aaron, he told me he thought the structure of GH5 wouldn't be that interesting by itself and suggested I do some experiments to get it to bind to a piece of DNA. I really wanted to learn crystallography, but being in awe of his reputation and too afraid to contradict him, I started working on his suggestion, sharing a lab bench with his postdoc, Wes Sundquist. After about a month, I realized that this was not something that could be done on the time scale of my sabbatical, if ever. With some trepidation, I told Aaron why I thought this was not a useful approach and that I wanted to focus on solving the two structures using the data I had brought with me. Somewhat to my surprise, he readily agreed, and I think it increased his respect for me. Wes was amused as he watched an experiment I had abandoned slowly dry out and gather dust during the rest of our stay there.

Even if he'd had the time, Aaron could not have taught me the nuts and bolts of how to solve a crystal structure because the computing software had changed a lot since his day. Luckily, the LMB was (and remains) one of the most collegial places I have ever seen. Both junior scientists like Paul McLaughlin and well-known crystallographers like Andrew Leslie and Phil Evans took me under their wing and taught me the practical ins and outs of the method. Pretty soon I was looking at a detailed map of S5 and building an atomic model of the protein.

The excitement of building a structure cannot be exaggerated. Until that point, the molecule is a black box; you know it exists and you know what it does, but not a lot more. Now, suddenly, it is as though the curtains have parted, and you see the molecule in its full glory, all of its atoms in their place revealing the twists and turns of the chain as it folds up into its unique architecture to show how it might work. It must be how explorers felt when they came across a completely new landscape.

Solving my other structure, GH5, had an interesting twist. Nearly every structure solved using MAD up to that point had used an elaborate formalism developed by Wayne Hendrickson. It used complicated bookkeeping to keep similar measurements made at different wavelengths separate and was cumbersome to use. More importantly, it didn't have the kind of sophisticated treatment of errors that standard crystallographic programs had. But we all thought that the special treatment was needed because the signal was so small. A typical gold or mercury heavy atom had about eighty electrons, but the typical variation in the properties of selenium from one wavelength to another in a MAD experiment was only a few electrons' worth. It was a miracle the method worked at all because the difference of a few electrons in scattering power was like adding a water molecule to the structure – hardly expected to make a difference.

Using Wayne's elaborate programs, I had obtained what I thought was a decent map of GH5 and started building a structure of it. One day, Phil Evans returned from a visit to York, where his friend Eleanor Dodson thought a new program for solving standard heavy-atom structures might also be useful for solving MAD structures. I was sceptical but gave it a shot. To my surprise, within a few hours, I had a much better map, which we used to solve and publish the structure of GH5. After our paper, virtually nobody used Wayne's programs any more and instead began solving MAD structures with the same software they used to solve their other structures.

If the signal from a MAD experiment is so small, why did the standard software work at all? I started thinking about it and realized – perhaps belatedly – that MAD worked even though the basic signal was small because the errors in the experiment were even smaller. What matters is not just the strength of the signal but how much larger it is than the error or 'noise' in the data, or what scientists call the signal-to-noise ratio. In the standard heavy-atom method, when you collect data on crystals with and without a heavy atom, the two crystals are never identical – they have different shapes and vary slightly. To get a proper difference between them, you have to make sure all these factors are compensated for so that the two sets of data are on the same scale, which is quite hard. Another problem is that adding a heavy atom changes the rest of the structure, a problem called non-isomorphism. This means you can never accurately compare the two data sets. In MAD, these problems don't exist. Data from the two wavelengths are collected on the same crystal, and the anomalous differences between symmetry-related spots are measured not only on the same crystal but at the same wavelength, often at the same time. So the signal-to-noise ratio in a MAD experiment is very good, and it typically produces much better maps that are easier to interpret in atomic terms.

The full implication of how powerful anomalous scattering was didn't sink in right away. I had no idea how important a role it would play in my future. At the time, I was just pleased not to have made a fool of myself in my sabbatical among towering figures like Aaron. On the contrary, I had accomplished exactly what I wanted during my year in Cambridge, and the two structures I solved were published in *Nature* soon afterwards. Moreover, I had made a lot of friends and connections. What I didn't quite realize then was that my sabbatical had changed forever both my attitude towards science and how I would approach the ribosome. When I returned, I found I was no longer satisfied with making solid but incremental advances and instead wanted to attack the big questions in the field.

CHAPTER 6

Emerging from the Primordial Mist

DURING MY SABBATICAL, TWO PAPERS made me rethink my approach to the ribosome. One of them dealt with a long-standing chicken or egg paradox about how the ribosome could even have emerged. All life today depends on the thousands of reactions carried out by proteins. The ribosome, the machine that makes proteins, itself consists of lots of proteins. So how could the ribosome have even come into existence? Once again, it was Crick who had a key insight. In a classic paper in 1968, he was struck by the fact that although ribosomes had lots of proteins, they were made up mainly of RNA. What was the purpose of this ribosomal RNA? Crick suggested that ribosomal RNA and tRNA '*were part of the primitive machinery* of protein synthesis' (italics in the original). He then went on to say, 'It is tempting to wonder if the primitive ribosome could have been made *entirely* of RNA.'

The problem was that when Crick proposed this, all known enzymes, the biological molecules in every cell that could carry out the chemical reactions that are essential for life, were proteins. Nucleic acids like DNA or RNA could at best be inert carriers of information. There was not a shred of evidence that RNA could carry out any kind of chemical reaction, let alone direct a complicated process like translating genes into proteins. By then, other enzymes involving DNA and RNA had been discovered – ones that made copies of DNA during cell division or others that copied DNA into messenger RNA. They were all made up of proteins.

So the general feeling was that ribosomal RNA was just a kind of scaffold on which to hang the various proteins, each of which would do one of the many jobs of the ribosome. One protein might help tRNA to recognize the code, another could help add amino acids to the growing protein chain, and so on. That explained why there were so many ribosomal proteins.

Early work on antibiotics supported the idea that it was the proteins that were performing the many important tasks in the ribosome. It had been known ever since the 1950s that many antibiotics worked by blocking the ribosome. Nomura showed that mutant bacteria that were resistant to streptomycin had an altered ribosomal protein. Mutations in other proteins also changed how the ribosome responded to antibiotics. If changes in proteins were making the ribosome behave differently, it must be because they were doing something important.

Although lots of scientists were focusing on the proteins, a few were intrigued by ribosomal RNA. One of them was Harry Noller, whom I would jokingly refer to as the Sage of Santa Cruz. With his long hair and beard, typically wearing jeans with a T-shirt, he superficially had the demeanour of a mellow, pot-smoking California hippie. He also had a fondness for motorcycles and antique Ferraris (even his computers were named after Italian racing-car drivers). His cool charisma and wry wit meant that in public gatherings he was sometimes surrounded by adoring young scientists, much like groupies around a rock star. He had a large cult-like following among his many protégés, who viewed him as the spiritual leader of the ribosome. But underneath it all, he was a serious and deeply ambitious man with a relentless focus on the ribosome.

A California native, Harry got his undergraduate degree at Berkeley before going on to get his doctorate in protein chemistry from Oregon. Then he too became a postdoctoral fellow at the LMB. There, under Ieuan Harris, he was working on a protein that is involved in metabolizing glucose. In an autobiographical essay, he describes how he was feeling slightly intimidated by the whole scene at a party in one of the Cambridge colleges, when Sydney Brenner came up to him and asked him who he was and what he was doing. On hearing that Harry was working on glyceraldehyde phosphate dehydrogenase, Brenner announced, 'That's stupid! If you're a protein chemist, why don't you work on something *interesting* like ribosomes?' In hindsight, this is quite ironic considering Brenner thought ribosomes were not interesting enough to pursue at the LMB himself.

Harry was initially crushed by this scathing assessment. But then, in a bold decision, he decided Brenner was right and left Cambridge to work with Alfred Tissières in Geneva. There, he overlapped slightly with Peter Moore, who was doing the same European tour in the opposite direction, going to Geneva first and then to Cambridge. Peter said that Harry was hired in Geneva for his protein expertise so they could purify and characterize all the ribosomal proteins. After all, that must be where the action was.

When Harry went back to California and started his own lab in Santa Cruz, he did a key experiment that changed the course of his life. He and his student Jonathan Chaires showed that if you modified the ribosomal RNA in the small subunit with a chemical called kethoxal, it could no longer bind tRNA. This was the first clue that ribosomal RNA actually might have an important function. Many scientists might have dismissed it as a curiosity, but to Harry's credit, he followed this up and became a lifelong RNA biologist.

Then in the early 1980s, there was a scientific earthquake initiated by two scientists, Tom Cech in Colorado and Sidney Altman at Yale. Cech was looking for the enzyme that carried out a reaction in which a stretch of RNA could excise itself from a longer piece. He found that just the RNA could excise itself without any help from a protein. Altman, on the other hand, was studying the properties of an enzyme that could cleave certain RNA molecules. The enzyme itself was a complex of a protein and RNA, and to his surprise, he found the RNA component by itself could carry out the cleavage reaction. So both groups had shown that RNA by itself could carry out a chemical reaction. Enzymes that were made of RNA came to be called ribozymes to distinguish them from the more usual protein enzymes. Although the reactions seem specialized, they have huge implications for the origin of life.

How life began is one of the great remaining mysteries of biology. All life requires some form of energy in the right chemical environment. Some people have pointed out that a lot of the chemistry that life uses resembles the kind of chemistry that occurs at the edges of geothermal vents in the ocean. Even if this is merely a coincidence as others have argued, it is useful to think about what conditions made it possible for life to emerge. But fundamentally life is more than a set of chemical reactions; it is the ability to store and reproduce genetic information in a way that allows complex life forms to evolve from very primitive ones. By this criterion, there is no question that even viruses are alive, even though people used to question it because they need a host cell to reproduce. However, anybody who has become ill from a virus and experienced his or her body fighting an infection would not doubt that viruses are alive.

The problem was that in nearly all forms of life, DNA carried genetic information, but DNA itself was inert and made by a large number of protein enzymes, which required not only RNA but also the ribosome to make those enzymes. Moreover, the sugar in DNA, deoxyribose, was made from ribose by a large complicated protein. Nobody could understand how the whole system could have started. Scientists who were thinking about how life began, like Crick, Leslie Orgel at the Salk Institute in La Jolla, and Carl Woese at the University of Illinois, suggested that maybe life began with RNA. At the time, this was pure speculation – almost science fiction – because RNA was not known to be capable of carrying out chemical reactions.

Cech's and Altman's discovery changed all that. RNA was now a molecule that could both carry information as a sequence of bases, just like DNA, and could also carry out chemical reactions like proteins. We now know that the building blocks of RNA can be made from simple chemicals that could have been around in the earth billions of years ago. So it is possible to imagine how life may have started with lots of randomly made RNA molecules until some of them could reproduce just themselves. Once this happened, evolution and natural selection could allow more and more complicated molecules to be made, eventually even something as complicated as a primordial ribosome. The idea of a primordial RNA world, a term first coined by Wally Gilbert, became widely accepted.

The ribosome may have started off in an RNA-dominated world but because it made proteins, it became a Trojan horse. Proteins turned out to be much better at carrying out most kinds of reactions than RNA because their amino acids are capable of more varied chemistry than the simpler RNA molecule. This meant that as proteins were made, they gradually evolved to take over most of the functions of the RNA molecules around at the time and much more. In doing so, they transformed life as we know it. This may also explain why although the ribosome has a lot of RNA, the enzymes that replicate DNA or copy it into RNA are now made entirely of proteins. This is probably because the use of DNA to store genes came later; by that time, proteins had become prevalent and were carrying out most of the reactions in the cell.

Of course, this doesn't explain how genes carrying a code to make proteins came into being. The best guess is that an early form of ribosomes just made short stretches of random peptides, which helped to improve the RNA enzymes that were around then. But from there, how genes originated that carried instructions to make proteins that had amino acids strung together in a very specific order was quite a leap and is still one of the great mysteries of life. And that in turn would mean that in addition to the large subunit, many other elements would have to come into existence: mRNA to carry the genetic code, tRNAs to bring in amino acids, and the small subunit to provide a platform for the mRNA and tRNAs to bind. But before the discovery of RNA catalysis, people couldn't see how the system could have begun even in principle.

Why could RNA carry out reactions and not DNA? The main difference between the two molecules is just an oxygen atom on the ribose sugar of RNA to form a hydroxyl (OH) group. We now know that this small difference allows the OH groups from different parts of the RNA molecule to bond with each other, so RNA can fold back on itself and have compact three-dimensional shapes like protein enzymes, with pockets to carry out chemical reactions.

After Cech's and Altman's discovery, everyone realized that Crick was probably right in suggesting that a primordial ribosome was *entirely made of RNA*. What about today's ribosome? Had key functions been taken over by proteins as they had with other enzymes? Or as now seemed completely plausible, did ribosomal RNA still carry out most of its important functions?

Meanwhile, Harry had continued his work on ribosomal RNA. He had no idea of the location of the chemical modifications that prevented tRNA from binding the ribosome. In fact at the time, nobody even knew the sequence of ribosomal RNA. Soon after Harry's initial results, Fred Sanger at the LMB had figured out how to sequence or determine the precise order of the bases in any given piece of DNA, for which he received his second Nobel Prize (he is one of very few people to have won two). So Harry returned briefly to Cambridge, this time to learn how to sequence DNA. Instead of trying to sequence the RNA directly, which remains a much more difficult task, Harry used Sanger's methods to determine the exact sequence of ribosomal RNA by sequencing their genes, which resided on DNA. The large ribosomal RNA from the 30S and 50S subunits are called 16S and 23S RNA, respectively.

Sequencing ribosomal RNAs turned out to be important by itself. By comparing sequences from different species, Carl Woese and Harry could work out how they were related and how the RNA molecule would fold back on itself so that parts of the molecule formed internal base pairs with other parts. This internal base pairing meant that ribosomal RNA had lots of segments that were double helical. Eventually, comparing ribosomal RNA sequences led Woese to discover that in addition to bacteria and eukaryotes, there was a distinct third domain of life called the archaea. It is now believed that an early bacterium joined an ancient archaeon to form the first eukaryote, whose cells have a nucleus. (Archaea, like bacteria, are prokaryotes, meaning they have no cell nucleus.) Eukaryotes then evolved into the complex multicellular organisms we have today, including humans.

Once the sequence of ribosomal RNA was known, Harry could then try to locate where on the RNA molecule his chemical agents modified it. He adapted a technique that people had developed to see where proteins would bind on DNA. Scientists would expose DNA with a chemical that modified it, then do the same when a protein was bound to it. The protein would protect those parts of the DNA that it bound because the chemical couldn't reach them. Afterwards, they could compare the modifications in the DNA with and without the protein, and the difference would tell them roughly where the protein sat on the DNA molecule, or its footprint. Harry and his students, most notably Danesh Moazed, started applying this method, called footprinting, to ribosomal RNA. They figured out which parts of ribosomal RNA were involved in binding to tRNA molecules and to each of the ribosomal proteins. This technique produced a lot of data on what touches what in the ribosome, but like other studies carried out in virtually every other lab in the world working on the ribosome, it did not produce a real sense of how it all fitted together, let alone how it functioned.

Doing footprinting with antibiotics was more interesting. Many antibiotics bind to ribosomes, but even though some altered proteins made ribosomes resistant to antibiotics, nobody could get antibiotics to bind to any of the ribosomal proteins by themselves. By footprinting, Harry was able to show that each antibiotic bound to a particular part of ribosomal RNA. Since antibiotics stop ribosomes from working, clearly ribosomal RNA must be doing something important. Because of Cech's and Altman's discovery that RNA could carry out reactions, and Harry's work on antibiotics, the field was quickly coming around to the view that ribosomal RNA played an important and possibly central role in the ribosome.

In any case, after being passé for a long time, ribosomes became interesting again. In a prophetic article in *Nature* called 'The ribosome returns,' Peter Moore wrote in 1988, 'Fashions come and go in biochemistry. The discovery that some RNAs are enzymes is reviving interest in the long-neglected ribosome.' Not even he could have predicted it would return with such a vengeance.

Towards the end of my sabbatical in 1992, a paper in *Science* by Harry generated a lot of excitement. He tried to settle the question of whether just the RNA part of the ribosome could carry out a key reaction of the ribosome called peptidyl transfer or the joining together of two amino acids to form a peptide bond. In other words, was the ribosome a ribozyme? He took 50S subunits from a species of *Thermus* bacteria that grow in the hot springs of Yellowstone National Park and treated them with an enzyme that digests and degrades the proteins into fragments. He then extracted as much of the remaining protein fragments as he could. The resulting subunits consisted almost entirely of RNA and could still carry out the reaction.

Harry's paper generated quite a buzz among scientists broadly, but by then the result was not a huge surprise to the ribosome community. It was also not definitive. Harry had gone to great trouble to chew away and remove the proteins in the 50S subunit, but there were still lots of protein fragments and even some whole proteins left in his particles. So the reaction could still have been carried out by a protein or a piece of one. When Harry used a different method to remove all the proteins completely, the particles were no longer active. The procedure didn't work with ribosomes from *E. coli*, which is what most people studied. Harry himself implicitly acknowledged the limitations of his work by his rather cautious title, 'Unusual resistance of peptidyl transferase to protein extraction procedures.' A few years later, in 1998, a group in Japan thought they had nailed it when they made pure fragments of ribosomal RNA that could carry out the reaction, but after publishing this with much fanfare, also in *Science*, they found flaws in their work and withdrew their paper a year later.

It was clear that after forty years of trying to solve how the ribosome worked by chemical methods alone, other means would be needed. In the same paper where Crick suggested that an early ribosome might have been made up entirely of RNA, he also said, 'Without a more detailed knowledge of the structure of present-day ribosomes it is difficult to make an informed guess.'

CHAPTER 7

A Threshold Is Crossed

APART FROM HARRY NOLLER'S PAPER in *Science*, the other paper that hit me during my sabbatical was a short report in 1991 by Ada Yonath on a dramatic improvement with one of her crystals of the large subunit. For the first time, at least in principle, the crystals were good enough to obtain an atomic structure of an entire subunit of the ribosome, with hundreds of thousands of atoms. A threshold had been crossed. To understand when a crystal is 'good enough' requires some explanation.

So far, we've assumed that crystals are formed by molecules packing identically into a three-dimensional stack, also called a lattice. This rarely happens with large molecules like proteins. During crystallization, a protein molecule finds its way into the growing crystal lattice but because it is large and floppy, it doesn't sit in *exactly* the same orientation as its neighbouring molecules. The final image is the result of adding up the contributions of the millions of individual molecules in the crystal. If the molecules are all sitting slightly differently in the crystal, their contributions will be blurred out to a greater or larger extent – think of what a multiple exposure photo of something immovable like a rock would be compared to a multiple exposure of a person who cannot stand perfectly still.

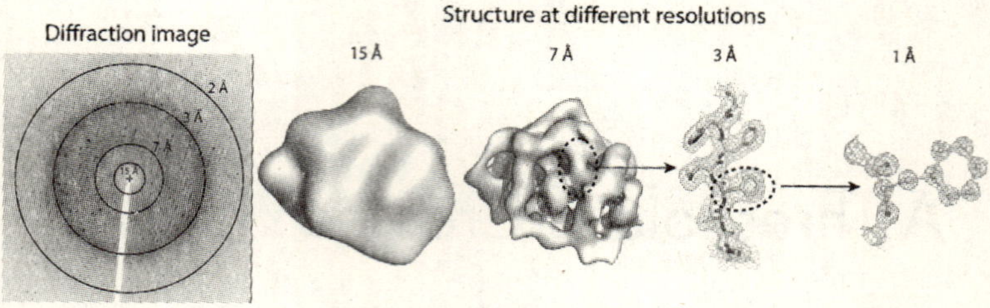

Figure 7.1 Diffraction and features seen at different resolutions

How good a crystal is depends not on how nice it looks visually but how well it diffracts X-rays. As I pointed out in chapter 3, resolution tells us how close two features can be and still be distinguished. In practice, you can tell how good the resolution of a crystal is by seeing how far out in angle the X-ray spots extend from the direction of the incident beam.

In the worst crystals there are only a few diffraction spots around the main X-ray beam. If you solved the structure from that sort of crystal, you'd only see a blobby-looking object with almost no detail, but it might give you a sense of its overall shape. For more moderate crystals, which yield data to, say, 5–7 Å resolution (Å represents the unit of Angstrom or 0.1 billionth of a metre), you could see some characteristic features of proteins and DNA or RNA. For example, you would just be able to see the grooves in the helices of DNA or RNA. The narrower single-chain alpha helices adopted by many proteins would look like tubes. In a well-diffracting crystal, the spots can be seen out to very high angles, all the way out to the theoretical limit imposed by the wavelength of the X-rays (remember you cannot see features that are closer than about half the wavelength). At that limit, when you obtain maps by solving the structure, you can see individual atoms as distinct spheres. This is often seen for simple molecules like salts, which have only a few atoms and have precisely the same orientation in a crystal. But even small proteins almost never reach 1 Å resolution like small molecules, and as the molecules get larger and floppier, it is very rare for them to produce crystals that diffract to very high resolution. In practice, this means the diffraction spots get weaker and weaker at higher angles, and then just disappear. So the information about fine detail is lost because there isn't enough order in the crystal to produce it. The higher the angle to which you can see diffraction spots, the higher the resolution of the structure you can get from the data, so crystallographers typically talk about the resolution of the crystal or even of a particular diffraction spot.

As the resolution improves beyond 3.5 Å, you are still nowhere near being able to see individual atoms (for which you need to be better than 1 Å). But you can still obtain an atomic structure because you can start to see the characteristic shapes of amino acids and bases. In proteins, amino acids vary in shape from large and flat, to long and thin, to short and stubby. If you know the order in which they should appear, then you can fit a protein chain with these amino acids into their shapes, much like solving a large three-dimensional jigsaw puzzle. Similarly, for DNA or RNA, the bases T (or U) and C are small, and the bases A and G are large. So even though you can't see the individual atoms, you can 'build' a chemical structure in atomic detail in these maps. If there are amino acids of very similar shapes near each other in the sequence, it is still possible to make mistakes because, unlike a jigsaw puzzle you buy in a store, the shapes in a three-dimensional map aren't perfect due to errors in the data. Also, unlike a jigsaw puzzle, this problem is in three rather than two dimensions, and there is no answer conveniently provided on the front of the box.

So the threshold of about 3.5 Å is the holy grail for people trying to get crystals: anything better means you have a good chance of solving an atomic structure, whereas worse than about 4 Å means that it won't be easy unless you already know roughly what the molecule looks like.

The original crystals of ribosomes were terrible – they produced hardly any diffraction spots at all. Rather than being discouraged, Ada Yonath, now working with her long-term associate Francois Franceschi who was overseeing the biochemical operations for crystallization in the Berlin institute, continued to toil away in a systematic effort to produce better crystals – even though there was no clear prospect of a structure from them. Ada was also carrying out some of this work in her lab at the Weizmann Institute in Israel. At the time, many scientists in Israel were identifying new species to characterize the local biodiversity, and had identified a microorganism called *Haloarcula marismortuii*, which, as its name *marismortuii* suggests (literally, 'sea dead'), grows in the extremely salty Dead Sea. Because it grew in a very high salt environment rather than very high temperature, it represented a different kind of extremophile – organisms that grow in extreme conditions. Later, it turned out that *Haloarcula* is not a bacterium but an archaeon – its ribosomes are more complex and in some ways halfway between those of a bacterium and a eukaryote. Ada thought its ribosomes might be worth trying to crystallize, and the large subunits from it yielded better quality crystals than any of the others. Initially, they were not good enough to produce an atomic structure, but by tinkering with the conditions in which the crystals grew, Ada's group was able to improve them.

Apart from getting good crystals, there was a major problem with actually collecting data from them: crystals of many proteins and especially of large molecules like the ribosome would be damaged by the X-rays themselves before the data could even be collected. These data are typically collected by rotating a crystal in a beam of X-rays and taking a series of snapshots to measure the scattered X-rays with a detector. For each orientation, some of the planes of the crystal will satisfy Bragg's law and produce spots in certain directions. When you collect all the possible spots this way, you have a complete data set and can start calculating your structure.

Ribosome crystals were particularly difficult to work with because the diffraction spots from them were so weak. This is because the larger the molecule, the fewer of them there are in a crystal of any given size. Since the strength of the spots is related both to the number of molecules (because they come from adding up the contribution of the individual molecules) as well as the intrinsic quality of the crystals, the diffraction spots from ribosome crystals faced a double handicap and were much weaker than those of typical protein crystals, let alone salt crystals. To see them at all, they have to be exposed to a very intense beam of X-rays for a long time. During this time, the X-rays damage the molecules and create disorder in the crystals. As a result, the molecules break up partially, and either their internal structure changes or they no longer sit precisely in the same orientation as they did originally. To make matters worse, the initial damage generates free radicals that spread throughout the crystal and cause even more damage. With a large crystal, you can often visually see where the X-rays have damaged the crystal because the free radicals generated by X-ray damage often have a faint but distinct colouration. The result is that the resolution of the crystal falls off with time during the experiment – the loss of resolution is directly visible because the diffraction spots at high angles fade away and disappear as the crystal is hit with X-rays.

Crystallographers think of this loss of order and resolution as the death of their protein crystal and talk about crystals 'dying' in the beam. With crystals of a smaller protein, you could collect all your data from one crystal, or if they started to die out with radiation damage, you could keep switching to new crystals until you had collected all your data. With ribosome crystals, you couldn't even collect your first diffraction image. Not only that, you had no way of knowing if the crystal had good enough resolution to begin with because, even before the first snapshot was completed, the spots might already have died out from radiation damage.

At some point, scientists realized that cooling crystals would slow down the diffusion of free radicals generated by X-rays and thus slow down the damage. The first real evidence that this worked was produced when David Haas, who had been a postdoc at the Weizmann Institute in Israel, joined Michael Rossmann in Purdue. Like many of his generation, Rossmann also got his start at the LMB, where he worked with Max Perutz on the first haemoglobin structures. From there, he went to Purdue, where he has spent the rest of his life and is now a world-famous scientist. Even now, in his eighties, he maintains an amazingly active group and is so energetic that he is known to climb up mountains faster than many young postdocs and students. Haas and Rossmann decided to cool crystals of an enzyme to minus 75 degrees C and noticed that the X-ray diffraction spots from protein crystals would die out from radiation damage much more slowly.

But simply cooling crystals to low temperatures would not work in most cases. Crystals are about half water, which is why even though they can often look very regular with well-defined faces and edges, they are squishy like jelly or crumbly like cheese if you poke them with a needle. Remember, proteins are irregular molecules, and when they stack up, they touch each other at only a few points, and there are large channels of water in between. So when you take a crystal and cool it to very low temperatures, the water in the channels freezes and expands and ends up basically destroying the order in the crystals. Greg Petsko, who was then at MIT, found a way around this by replacing the aqueous solvent in the crystals with something like an antifreeze. In some cases, this would still preserve the order in the crystals. This allowed him to collect data at low temperatures and, as Rossmann had noticed earlier, with very little of the radiation damage that you would see at room temperature.

For some reason these methods didn't catch on until later, perhaps because they were quite difficult to generalize, so if people didn't absolutely need them they just didn't bother. Meanwhile, electron microscopists, whose samples were also being damaged by irradiation with electrons, became interested in looking at samples at low temperatures too. Jacques Dubochet, who was working at the EMBL in Heidelberg, figured out that if he rapidly plunged his samples into liquid ethane, the water would cool so fast that there was no time for the water molecules to freeze by arranging themselves as ice. The water would vitrify or become like a glass, preserving the biological molecules in their native state.

Meanwhile, a Norwegian from UC Davis named Håkon Hope had been working on collecting data on several small organic

Figure 7.2 Joel Sussman, Felix Frolow, and Håkon Hope doing an early crystal cooling experiment at the Weizmann Institute *(courtesy of Joel Sussman)*

molecules that were sensitive to oxidation at room temperature. Perhaps influenced by Dubochet and others who were doing electron microscopy, he got the idea of coating his crystals with oil and then plunging them rapidly into liquid propane. The crystals maintained their order as he had hoped. Although we now routinely refer to the process as 'freezing' and the crystals as 'frozen,' it is important to remember that the method works precisely because the water doesn't freeze but vitrifies.

Håkon visited the Weizmann Institute in Israel where he met Joel Sussman, who asked him if he could try to adapt his method for biological molecules. Håkon went back and used this approach on two very small proteins. He then returned to the Weizmann Institute and worked with Joel and his associate Felix Frolow on making the method more general. One of the first crystals they tried was a crystal of DNA that Joel's student Leemor Joshua-Tor was working on. After that initial success, they got it to work on several other projects.

Although by then Ada had a large operation in Germany, she continued to run a lab at the Weizmann Institute. At some point, Håkon and others in Joel's lab encouraged Ada to try his method of cooling crystals to help her with the problems of collecting data from her ribosome crystals. She was initially sceptical but decided to give it a shot. It was not quite so straightforward to do it with ribosome crystals because they needed intense X-rays from a synchrotron, and cooling equipment was not standard on synchrotron beamlines at that time. Eventually, Ada and Håkon got some time to collect data at the Stanford Synchrotron Radiation Lightsource, which is only a couple of hours away from Håkon's lab in Davis. Håkon put his cooling equipment in his car and drove down to Stanford and set it up. According to him, the first freezing experiment worked very well, and they could see a nice diffraction pattern with lots of spots. The next dozen or so tries failed, but that first successful attempt gave them the confidence that it could work, so they kept going until eventually the technique worked more reliably. Once the method worked for ribosome crystals, Ada became an evangelist for the cooling technique, which became known as cryocrystallography.

Despite their success, the method took some time to be widely adopted because their original method involved painstakingly sandwiching a crystal that is typically less than a tenth of a millimetre long between two tiny shards of quartz that were glued onto the end of a pin. A few years later, in 1990, Tsu-Yi Teng at Cornell invented a simple procedure that involved fishing out a crystal from its drop with a tiny flexible loop at the end of a pin, where it was held in the loop by surface tension.

This turned out to be much simpler even than the old method of mounting crystals in narrow capillary tubes and led to a nearly universal adoption of cryocrystallography. It is often the case in science that just showing something works better is not good enough – it also has to be easy to use.

The original crystals of the 50S subunit from *H. marismortuii* were noticeably better than any that had been obtained, and after

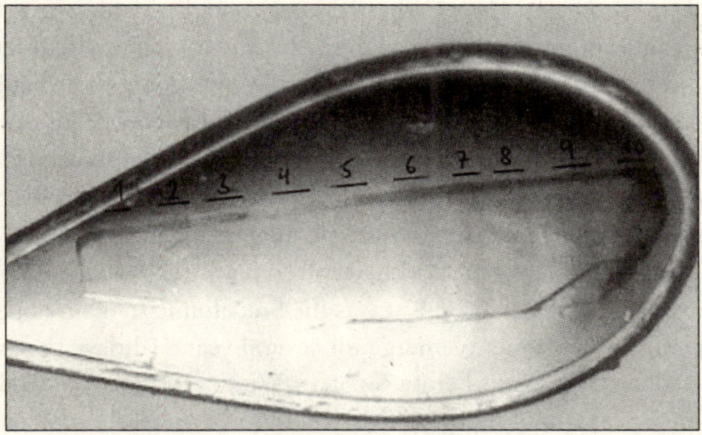

Figure 7.3 Crystal of a 30S ribosomal subunit cooled in a loop. The striations show regions that have been exposed to the X-ray beam. The crystal is about 0.3 mm long and less than 0.1 mm wide.

systematically tweaking the conditions to improve the quality of the crystals and cooling the crystals, Ada and her colleagues could now see diffraction spots out to 3 Å. This was well above the threshold needed to build an atomic structure. When the results were reported in the *Journal of Molecular Biology*, I had just begun my sabbatical and instantly recognized it as a major landmark. It was exciting to imagine that an atomic structure of the large subunit was now possible.

Set against this was the fact that around 1991, crystals of ribosomal subunits and even the entire ribosome had already existed for several years, and even if they didn't diffract to such high resolution, they should still have been able to yield maps that were detailed enough at least to see roughly where the individual proteins and RNA were. And yet, there were no such maps. Perhaps the ribosome was just too large for crystallography. Still, the paper by Ada's group in 1991 got me thinking during my sabbatical. I started wondering what she would do now that she had crystals that diffracted so well.

I was soon about to see where things stood. Every few years, the ribosome community meets in different parts of the world to catch up on the field. The next of these meetings was going to be held in Berlin soon after my sabbatical. It was a sad occasion, because Wittmann, who had done so much to make Berlin a centre for ribosome research, had died suddenly the previous year, and his colleague Knud Nierhaus had taken over organizing the conference.

Although the protein S5 I had solved on my sabbatical was just a tiny piece of the ribosome, it was the first atomic structure of any part of the ribosome to emerge for several years. I'd never been to one of these meetings before, so Steve White generously agreed to let me speak about it. There was nothing I particularly remember from the meeting, and certainly there seemed to have been no major progress on the structure of whole ribosomes beyond what I'd already read.

On my return to Brookhaven from my sabbatical, I was vaguely unsatisfied. After a year at the LMB, the contrast was striking. Oddly, in some ways Brookhaven was a lot like the LMB: scientists worked in small groups and did experiments themselves instead of managing people, and there was steady funding from the Department of Energy (DOE). But the bureaucrats at DOE had little sympathy for the small-scale science that produced real breakthroughs in biology. If they were scientists at all, they generally came from a physics background and saw national labs as a place for large facilities, like big particle accelerators or reactors. As a result, I could see that the biology department, which had produced some top-quality science, was being slowly squeezed for funds, which made it difficult to attract the fresh young blood needed to maintain the vitality of an institution.

A few months after I returned, I wrote to Richard Henderson, who was then head of the Structural Studies Division at the LMB, saying I'd enjoyed my sabbatical so much that I wondered if they had a permanent job for a guy like me. He replied that they had all enjoyed having me there, and while they all thought it would be nice to have me as a colleague, they didn't have any space or positions open, but I should keep in touch. I took this as a polite rejection.

Around this time, Wes Sundquist, with whom I'd shared a lab bench on my sabbatical in Cambridge, invited me to visit Salt Lake City, where he had just joined the faculty at the University of Utah. They had a mixture of young enthusiastic faculty and some well-known older ones, all in a beautiful setting surrounded by mountains. So when they asked me after my visit whether I'd be interested in a faculty job there, I was pretty excited about the possibility.

Suddenly, two other positions opened up at the same time, but given both the people and the setting, I felt Utah would be the right place for me. They made me an offer that was 50 per cent more than I was making at Brookhaven, and it made me a little uneasy because I thought I couldn't possibly justify that kind of salary. Soon after I said yes, all my fears about funding kicked in. At Brookhaven, even if you didn't get any outside grants, they would pay your salary and give you enough money to do science with a technician or two. At a university, you would be dependent on getting federal grants, and I had nightmares of losing my grants and having my career go downhill. So I called up Dana Carroll, the chair in Utah who had been extremely warm and friendly, to say that I was sorry I wasn't coming after all. He was not happy.

Soon afterwards, the starkness of staying on at Brookhaven also sank in. Apart from the problem of DOE slowly strangling the department, neither Vera nor I ever really liked living on Long Island, having to drive around everywhere and dealing with hot muggy summers and cold damp winters, which made my asthma worse. Rather contritely, I called Dana again and asked, 'Could I change my mind again?' He graciously agreed, but pointedly said he wasn't 'going to jump up and down this time.'

One reason for my initial hesitancy was that a germ of an idea was slowly forming in my mind but seemed so risky that I wanted to be secure enough in my personal situation to have the guts to tackle it. During my sabbatical, I had used the MAD method to solve the small GH5 protein and had been surprised by how such a small anomalous scattering signal from a selenium atom could give such beautiful maps. Could you use MAD to solve something really huge like a ribosome? It turned out that ribosomes don't have a lot of methionine residues, so the signal would have been too low.

But Wayne Hendrickson, who had pioneered using selenomethionine, had also used a different kind of atom for one of his structures. When he used holmium atoms to solve a protein structure using this method, his maps were not just good but spectacular. The reason is that holmium and other lanthanide elements have a much larger anomalous scattering at particular wavelengths. Could you use one of these lanthanides to solve the ribosome? When I did the calculation, it turned out that you would only need about a dozen of these atoms bound to a ribosomal subunit to get the kind of signal you'd get from a typical protein that had already been solved using MAD. I already knew that many of these lanthanide atoms bound to ribosomes in a dozen places because a recent paper showed exactly that.

I could hardly contain myself. This might be the secret bullet that would give me a route to solving the structure of something as large as a ribosomal subunit or even the entire ribosome. I redid the calculation several times to make sure I wasn't fooling myself, but the answer was always the same. If I had decent crystals, just a dozen or so of these metal atoms bound to a ribosomal subunit should give me its structure.

When I was mulling over this idea, I remembered my encounter with Ada's lieutenant Francois Franceschi, a Venezuelan of Corsican descent, whom I met for the first time at the conference in Berlin. Francois was very friendly and took Steve White and me to see his lab in Wittmann's old department at the Max Planck institute in Dahlem, a posh area of West Berlin with lots of famous scientific institutes that predate the war. It was the same institute where Steve had worked before coming to Brookhaven. During our chat, Francois told me that their work was reviewed by a committee every few years. The latest feedback was that they were spreading their efforts too thin and should focus on the 50S subunit, now that they had good crystals of it.

As I remembered this conversation, I realized that this meant that nobody would be focusing on the small subunit or the whole ribosome, neither of which had yielded good crystals so far. I thought the whole ribosome was premature to tackle, but the small 30S subunit – which binds the mRNA and helps to read the genetic code – was only half the size of the 50S subunit and seemed like a better bet. Suddenly I sensed an opportunity to enter the big leagues. But it would have to wait just a bit longer because I was about to move to Utah after twelve years in Brookhaven. In the meantime, there was another ribosome meeting to attend.

CHAPTER 8

The Race Begins

THE YEAR 1995 WAS A watershed for both the ribosome and me. I was planning to move to Utah that autumn, so I decided to stop off in Salt Lake City on my way to Victoria, the capital of British Columbia, where the next ribosome meeting was going to be held.

Vera and I spent our stay looking at houses in Salt Lake City and settled on one in the foothills with a spectacular view of the valley. We then went backpacking for a few days in the Hoh Rain Forest on the Olympic Peninsula and took a ferry from there to Victoria, which was just across a strait at the southern tip of Vancouver Island. Victoria is a charming and picturesque city that appears to relish its colonial past, evident in its British architecture and landscaping. Right in the middle of our meeting, a large parade celebrating the birthday of Queen Victoria marched by outside.

At the meeting itself, there was one exciting development and a surprising disappointment. The exciting development was seeing three-dimensional images of the ribosome from electron microscopy. The method had long been used to look at ribosomes and deduce their overall shape. But recently, researchers had started applying a method called single particle reconstruction to objects like ribosomes that had no symmetry; previously, the method had only been used to study regular objects like viruses. Biological molecules have very low contrast – their scattering for X-rays or electrons is very similar to that of the water in which they exist. So previous work in electron microscopy had involved coating the particles with a heavy atom like uranium and looking at it in an essentially dry state. At best you saw its surface shape, possibly distorted by drying out, and the level of detail was quite low. Despite the low contrast without stain, if you could extract enough signal from the particles, you might be able to get at the internal structure by electron microscopy. Richard Henderson and Nigel Unwin had done this with their two-dimensional crystals. It wasn't clear whether you could extract enough signal from individual asymmetric particles, but as a result of work by Jacques Dubochet, you could look at biological molecules at very low temperatures by plunging the samples rapidly into liquid ethane. Just as with X-rays, the damage in the electron microscope is slowed down at low temperatures. You could expose the samples to a higher dose of electrons and see individual biological molecules even without stain.

One of the leaders in the field was Joachim Frank, a German scientist who had been working away in relative isolation at the Wadsworth Center in Albany for years, where his lab was in the basement of a large government building. A tall, courteous, and somewhat reserved person with a deep love of art and literature (he even likes to write fiction and poetry), he also seemed quite insecure, perhaps because, like me, he had spent much of his career outside the fast track of the big important centres of science. He was developing methods to extract a meaningful signal from the noisy images produced by biological molecules.

At one point around 1980, he was joined by the Dutch microscopist Marin van Heel, who with his outgoing and brash manner was a striking contrast to Joachim. Soon after Marin went to work with Joachim in Albany, it became clear the town was too small for the two of them because he did not stay very long. Regardless of any other factors, their markedly different personalities would not have been a recipe for a prolonged and friendly scientific collaboration in any case. Marin ended up being recruited to Wittmann's institute in Berlin but during his short stay in Albany, he and Joachim managed to write a key paper that showed how to extract information by taking the noisy two-dimensional projections of particles in a microscope image and classifying them into groups that corresponded to different views of the particles. From that breakthrough, they both went on separately to develop methods to obtain a three-dimensional structure.

The images of the ribosome that Joachim showed at the meeting were the most detailed we had ever seen. For the first time, we could directly see the tRNAs nestled between the subunits and the mRNA snake around a cleft in the small subunit. They were still at far too low a resolution to deduce atomic structures or even visualize an arrangement of the various proteins and RNA components of the ribosome. They looked like a collection of blobs that made up an overall shape – so much so that those of us who did crystallography and were proud of being able to get atomic structures would scathingly refer to the method as 'blobology.'

By contrast, progress on the crystallography was disappointing. Many of us wondered what Ada had done with her very good crystals of the large subunit that she had produced five years earlier. She reported that she had finally obtained phases with heavy atoms for her crystals and had maps that were at 7 Å resolution. At this resolution, the grooves for the double helices of RNA should become visible. Individual proteins, especially those that we had already solved separately, also become recognizable, and their structural elements like helices can be seen. But Ada's three-dimensional maps seemed to have no recognizable features at all.

The audience, which consisted mainly of ribosome biochemists and geneticists, had no idea what to make of it. These kinds of specialized meetings are useful not only because we hear about the latest cutting-edge work, but they are where scientists raise their concerns and debate with one another. This sort of frank give-and-take – even if heated at times – is what helps science move forward.

So when the time came for questions, I raised my hand and asked, 'I know of at least two cases of important structures at 7 Å resolution. The first is bacteriorhodopsin in which the helices of the proteins could be seen as tubes, and the second is the nucleosome in which you could clearly see the grooves of the DNA double helix. Since we know the ribosome has both, why aren't you seeing them in your maps?' Ada replied, 'Well, if the ribosome were as simple as bacteriorhodopsin, it would have been solved a long time ago. And moreover, do you even know what the first maps of the nucleosome looked like?'

The features we should see at a given resolution should not depend on the size or complexity of the object, but having made my point, I didn't want to prolong the argument. After the session, I was standing around with Peter Moore, Harry Noller, and a few others. Harry looked very contemplative and asked what we thought the problem was. We agreed that something appeared to be seriously wrong.

The proceedings of the meetings are usually written up in a book, and Ada's chapter had the grandiose title 'A milestone in ribosome crystallography.' I thought it was closer to a millstone. At least one part of it turned out to be wrong – the way in which the molecules were arranged in the crystal lattice, a knowledge of which is a prerequisite to be able to calculate phases and the structure. We didn't know it at the time, but in hindsight, this error may have been one of the reasons Ada's maps in Victoria were uninformative.

I went away from the meeting thinking that rather than just contributing to the effort, I had a chance to make the next real breakthrough. Without my realizing it, Peter and Harry had come to the same conclusion. None of us said anything to one another at the meeting, but we all went back and began working on different aspects of ribosome crystallography. It was like the opening scene from the movie *It's a Mad, Mad, Mad, Mad World* in which just after a car crash, an old man tells a group of assembled strangers that he buried the loot from a robbery in a park. The people all pretend that they don't believe him but very quickly rush off to get a head start on the others.

Any remaining reservations about tackling the problem were banished a year later during a meeting of the International Union of Crystallography in Seattle. Unlike the ribosome meetings, which brought together everyone who studied the ribosome regardless of the technique they used, this meeting brought together people who used crystallography regardless of what they were studying. I was invited to give a talk based on the work I had done on obtaining phases using the MAD method. It was the first time I had been asked to speak at an international meeting on crystallography, and as a relative newcomer to the method, I was naturally pleased. But another important reason to go was that Ada was going to be there. There were widespread reports that she now had crystals of the entire ribosome that diffracted to 2 Å resolution, which was astonishing. If true, it meant that her crystals were now so good that it would be quite difficult for us to compete with her.

I was a little worried about having to speak to an audience of 'real' crystallographers. These people had actually developed the method; I just used it. Fortunately, my talk went off reasonably well. Ada was scheduled to speak during a Monday morning session called 'Macromolecular Assemblies,' which included a number of exciting talks about large, challenging structures. It was chaired by Wim Hol, a well-known Dutch scientist who had moved to Seattle recently. Each speaker was allotted about twenty to twenty-five minutes, with about five minutes for questions. The main job of the chair was to make sure that the speakers kept to the time and that the questions weren't dominated by just one or two people.

Ada began with a short history of her work and then described her new crystals of the entire ribosome. She went into great detail about how she characterized them and why she believed they were ribosomes. She mentioned that they diffracted to 2 Å resolution. The diffraction pattern she showed was stunning, with lots of spots going out all the way to the edge of the photograph. We were hanging on to the edge of our seats waiting to find out what she had done with them.

She spent most of her talk describing how she tried to further characterize them before hitting us with the anticlimax: they weren't ribosomes at all but a contaminant, which turned out to be a protein called enolase! By this time, she had used far more than her allotted time, but Wim Hol was unable to stop her. She went on to describe her progress with real ribosome crystals. As far as we could tell, there had been hardly any since the previous year in Victoria.

Several of us in the audience could not understand why Ada spent so much time talking about a contaminant. By the time she finished, the meeting was running over half an hour late. This had an unfortunate consequence for Paul Sigler from Yale, the last speaker in the session. Paul was a giant, both physically and scientifically. He too had got his start at the LMB and had solved a large number of very important structures during his career, becoming one of the most prominent structural biologists of his generation. A brash and confident man, he was known to be plain spoken with a very hot temper. He once broke the glass panel on a Xerox machine by thumping his hand down on it in anger; on another occasion, he had smashed a desk drawer. But he was widely respected and admired.

Paul had deliberately been put at the end of the programme to cap the session because his lab had just solved the structure of GroEL, a large protein complex that helps newly made proteins fold up properly after they have emerged from the ribosome. But the moment he got on the podium to speak, some of the stage crew came and hustled him off, saying, 'Sir, you have to stop and leave immediately. We have a Nobel laureate speaking here in just a few minutes.' Paul was furious. Apparently, the conference had organized a lecture to be given by a different Nobel laureate at lunchtime, and that day, it was the very same Hans Deisenhofer who, a few years earlier, had lectured in the Cold Spring Harbor course I had taken just two days before he got his Nobel Prize. All of us liked

Deisenhofer, but Nobel or not, none of us thought that he was in any way more worthy to speak than Paul.

The buzz at the meeting was that Tom Steitz, a well-known crystallographer and colleague of Peter Moore's at Yale, had started working with Peter on ribosome structure. Tom had gone to Lawrence University in his home state of Wisconsin before doing his PhD at Harvard with Bill Lipscomb, a famous chemist who was also one of the early American protein crystallographers. There he met his wife, Joan, who was working in Jim Watson's lab on ribosomes.

Joan, today one of the leading molecular biologists in the world, first tried to work with a well-known cell biologist at Harvard for her PhD, but he refused to take her, saying, 'But you're a woman. What are you going to do when you get married and have kids?' She could barely contain herself and burst into tears as soon as she had left his office. Fortunately, Watson had no hesitation in taking her on, and she was in his lab at a particularly exciting time in the very earliest days of ribosome research.

Tom and Joan went to the LMB for their postdoctoral work, a common theme for many characters in this book. There, Tom worked with the leading crystallographer David Blow on chymotrypsin, a protease or an enzyme that cuts other proteins. Joan had a ruder introduction to the LMB. Watson had written to Crick asking him to find a place for Joan, but on her arrival at the LMB, Crick told her there was no space for her and she could spend her time doing 'library research'! Luckily, Mark Bretscher shared some of his space with her, and Joan went on to make the hugely important discovery of how ribosomes get started at the right position on mRNA. A lot of socializing among academics happened during dinners at High Table of the various colleges, so called because it is a long dining table on a slightly raised platform where only fellows and their guests are allowed to dine. Mark thought it was high time that women were included in this activity. At the time, most Cambridge colleges would not accept women as fellows (the only exceptions being women's colleges such as Girton or Newnham). To get around this restriction, Mark proposed Joan as a 'Member of the Room' at Gonville and Caius College where he was a fellow. It was the first time a woman had the privilege of dining there without being someone's guest.

By the end of their postdoc, Tom already had an offer from Berkeley, but on the way back, he and Joan had interviews at Princeton and Yale, and both universities gave both of them job offers as they walked out the door. As Tom describes it, 'When I got to Berkeley, I put the four letters on the desk of the chair of biochemistry when I went to visit him in his office. I asked if there was any possibility of a job for Joan at Berkeley. He looked at the letters and looked at me and said, "She's a woman. Women do not run their own lab; they work in the lab of their husband." We went to Yale.'

They have been a star couple at Yale ever since, each excelling in his or her field. Joan pioneered many areas of molecular biology, including discovering the molecules called spliceosomes that chop and splice RNA in higher organisms before it is read by the ribosome. She would often get honours before Tom, like being elected to the National Academy of Sciences, and it is still a source of mystery to some of us that she hasn't yet been awarded a Nobel Prize for her work. For a long time, many of us thought she was on track to get it before Tom.

When I first encountered Tom at Yale, he had already become one of the leading crystallographers of his generation. He was a burly, imposing figure who worked out at the gym with his friend and colleague Don Engelman. They both sported chin-strap beards, which made Tom look a bit like an Amish. Initially, Tom struck me as arrogant and pompous. But this was just my own deep insecurity showing, because later I realized he simply liked to be very direct – a characteristic of Midwesterners that I am intimately familiar with since I have been married to one for decades. Over the years we became good friends, and even though we ended up being rivals of a sort, he was always complimentary and supportive of me. But his directness, especially when combined with his enviable success in determining the structure of one incredibly important protein after

Figure 8.1 Tom and Joan Steitz in 1978 *(courtesy of Cold Spring Harbor Laboratory)*

another, did not endear him to some. Although many of us would marvel at his superb judgement and track record, occasionally I would hear the snide comment, 'Well, you'd have great judgement too if you were married to a top molecular biologist.' Being married to Joan may have helped, but those of us who knew Tom also knew he had a clear vision of trying to understand the underlying mechanism of the central dogma, or how information in DNA is replicated, copied into RNA, and used to make proteins. He had tackled nearly every aspect of the problem, even though some of his projects took years of persistence. So it was hardly surprising that he would want to tackle the ribosome.

Learning that Tom and Peter were going to team up to work on the ribosome initially caused me some serious anxiety because I was sure, with their combined expertise in the ribosome and crystallography, they would make formidable competitors. I was relieved to find out that they were focusing entirely on the 50S subunit, taking Ada's improved crystals from *H. marismortuii* as a starting point. I ran into Tom at the meeting in Seattle and asked him if it was true that he was going to go head-to-head with Ada. He smiled and said, 'Well, we hope to diverge at some point!'

I too had very briefly considered using Ada's 50S crystals to try my phasing strategy but didn't want to go into competition with her, partly because I thought people would not take kindly to my jumping on her problem. Of course, nobody 'owns' a problem in science. Once results are published, anyone is free to follow up on them and even compete with the original discoverers. That right of anyone to pursue anything worthwhile regardless of who began it or who else is doing it has been very beneficial for science. But in the early days of protein crystallography, when both X-ray equipment and computers were quite primitive and it took such a long time to collect data and solve a structure, there was a tradition that if someone had produced crystals of something, they were usually left alone to solve the problem. After all, there were plenty of other protein structures to solve. But the ribosome was different because it was so important and so many years had gone by since crystals had been obtained without much apparent progress towards an actual structure.

This was despite the fact that, right from the outset, Ada had strong support not only from Heinz-Günter Wittmann but also from people like John Kendrew, who saw similarities between her goal of the ribosome structure and his own work on the first protein structure. She had a purpose-built lab near a synchrotron with lots of support from the Max Planck Society, including the operation in Berlin to make ribosomes and produce crystals. The Weizmann Institute had given her an exceptional long-term permission to work in Germany while allowing her to keep her tenured job and lab in Israel. The broader community, too, was appreciative: right from the first breakthrough, she was regularly invited to speak at major international meetings.

However, neither she nor anyone in her lab had solved large, difficult structures. Nor was she collaborating with an experienced crystallographic group. So to me, it was as if someone who had never climbed a single major peak had decided to lead the first expedition up Everest without taking along an experienced sherpa. The same bold stubbornness that allowed her to take on what appeared a nearly impossible problem in Wittmann's department, and persist for years to finally get good crystals of the 50S subunit, was now proving to be a hindrance in getting to the next stage. It was not surprising that fifteen years after the first crystals were reported, even those who admired her efforts were getting frustrated and impatient.

Still, now that Tom and Peter had gone public, I wondered how they would be perceived by the community. At a dinner during the meeting, I asked the well-known British crystallographer Guy Dodson what he thought of Tom using the crystals Ada had originally discovered to compete with her. He replied without the slightest hesitation that it was about time others started working on it.

In any case, even though I had a small group with limited resources, I felt I had a chance to be part of the action. At least I wouldn't be competing with anyone, I thought, and certainly not with my old mentor, Peter, and Tom. They could duke it out with Ada in a race for the 50S subunit while I quietly carved out my niche by working on the 30S subunit. I returned to Utah with a renewed determination to crack the problem.

CHAPTER 9

Getting Started in Utah

STARTING ALL OVER AGAIN IN Utah was depressing at first. In Brookhaven, I had left behind a well-equipped working lab with a couple of excellent associates who understood the work and lots of colleagues who knew and liked me, had helped me grow into an independent scientist, and had taught me crystallography and the tools of molecular biology. For quite a while, I was alone in the lab. I was used to working on my own, and for several months, students and postdocs would pass by the hall and see the curious sight of a lonely full professor unpacking boxes and setting up equipment and trying to get some work done.

However, people in Utah were very welcoming and supportive, and within a year I had assembled a group working on the ribosome that consisted of two graduate students, a postdoc, and a technician. In addition, there were others who worked on chromatin, including Bob Dutnall. It was a motley crew. Our bread-and-butter project was still the work for my grant from the National Institutes of Health (NIH) with Steve White to bang out structures of individual proteins in the ribosome. Since there were a lot of proteins, it might have kept us going for a while. But by then, I was totally convinced it had become a protracted stamp-collecting exercise, and my heart wasn't in it even though I tried to talk it up. But it was a good way to train new people in crystallography by starting them on a project that wasn't too difficult.

The first to join was Bil Clemons, a big, tall African American with short hair and large glasses, who insisted on spelling his first name with just one *l* to distinguish himself from his father. He came from a highly accomplished family: his father was a US Marine Corps captain where he was the conductor of their band, and his uncle Clarence was the saxophonist in Bruce Springsteen's E Street Band, a role now taken up after Clarence's death by Bil's brother Jake. Bil arrived with a combination of enthusiasm and naïveté and a touch of immaturity. At some point after he joined my lab, he stopped cutting his hair, braiding it into dreadlocks as it grew longer and acquiring a distinctly Rastafarian look. He was extremely social, with a fondness for barbecues, beer, and hip-hop music. As a vegetarian and near teetotaller who liked mainly classical music, I was about as far from him culturally as one could get. But somehow, we hit it off.

Bil initially tried out my lab by doing one of the first-year rotation projects for graduate students. During his short stint, he had managed to get crystals of S15, a small ribosomal protein. At the end of the year, he had to decide where he wanted to spend his next few years for a PhD. At the time, my nearly empty lab might not have seemed very attractive compared to the thriving labs of Wes Sundquist and Chris Hill next door. But with S15, he already had a working project and that helped to sway him.

We had a chat, and I told him that he could start working on this protein, but really I had my sights on the whole 30S subunit. I thought he'd be sceptical, especially on learning that Ada's large well-funded group in Germany had been unable to crack its structure even after many years. To my pleasant surprise, his eyes lit up. I thought it was best for him to tackle a couple of smaller proteins first so he could learn the basics of solving structures before going on to the 30S. Bil's energy and optimism, and his willingness to try new non-standard approaches, were all invaluable in the years to follow.

The next to join was Brian Wimberly. He had called me about joining my lab in Brookhaven, but I told him I was moving to Utah, so he waited and contacted me a year later. He had obtained his PhD from Berkeley under Ignacio 'Nacho' Tinoco, where he was using a rather different method called NMR (nuclear magnetic resonance) to show that a piece of RNA in the ribosome had an unusual structure. He had worked on calcium-binding proteins during his postdoctoral work at the Scripps Research Institute in La Jolla, but never quite got over his romance with RNA. Towards the end of his postdoc, he had to decide whether to take the unusual – possibly foolhardy – step of turning down a faculty job at Georgia Tech or do a second postdoctoral stint with me to learn crystallography. I was eager to recruit Brian, who had real expertise in RNA structure, something I knew little about even though the ribosome was two-thirds RNA.

Brian came to visit me, and we hit it off right away. I had told him what a great and safe place Salt Lake City was, but when we emerged from my house the next morning, someone had smashed the windows of every car down the block. I was worried that this would derail my efforts to recruit him, but luckily, Vera took him for a hike up the foothills surrounding Salt Lake City where they saw lots of wildlife on a beautiful spring morning. Since he loved hiking and the outdoors, that sealed the deal for him.

The very next year, John McCutcheon, a smart, cocky, and ambitious student from Wisconsin, wanted to join the lab. I thought he could start off by working with Brian to get crystals of a ribosomal protein that bound to a small defined piece of RNA, but he soon felt this was too pedestrian and quickly decided to go for broke by working on the 30S structure. So now I had two students committed to working on 30S. Clearly, the fact that neither Bil nor John knew much about either crystallography or the ribosome had made it easier to convince them that this was a good thesis project!

Figure 9.1 The author's lab in Utah, with the author, Joanna May, Bob Dutnall, Brian Wimberly, John McCutcheon, and Bil Clemons *(courtesy of Isao Tanaka)*

The problem was how to begin. The crystals of the 30S subunit first reported by Garber's group in Russia were from *Thermus thermophilus*. However, almost a decade had passed since those crystals were reported, and they still did not diffract well enough to give an atomic structure.

In considering how to improve those crystals, I remembered Joachim Frank's talk in Victoria. He had not only shown images of the whole ribosome with tRNA, but had also shown that the 30S subunit had slightly different shapes depending on whether it was by itself or part of the whole ribosome, suggesting it was a rather flexible molecule. The subunit was described in anthropomorphic terms as having a 'head' that is connected to the rest of the 'body' by a slender 'neck,' and the head was tilted slightly differently in the different structures. It was as if the head could wobble. Large molecules often have to be flexible to work – the movement of the head turns out to be crucial to allow tRNAs to move through the ribosome. But flexibility is terrible for getting good crystals, which depends on all the molecules being exactly the same and sitting in the same way in the crystal lattice. Perhaps, I thought, this floppiness of the head was why the crystals of the 30S subunit weren't good enough. If so, the solution would be to somehow fix the head so it couldn't move about.

There is a protein that helps the ribosome get started at the right position on the mRNA called Initiation Factor 3 or IF3, and it was thought to bind right between the head and the body of the molecule. We had just solved its structure during my last year at Brookhaven, so it was on my mind. I suggested to John McCutcheon that he try to lock the 30S subunit by binding IF3 to it and then try to crystallize that.

With Bob Dutnall's help, we used a chromatographic column to purify away the 30S subunits from any contaminants, especially enzymes called ribonucleases or proteases that could destroy the ribosome by degrading its RNA or protein. This was important for two reasons: it meant the 30S subunits could sit in drops waiting to crystallize for weeks and remain intact, but it also had the effect of removing one ribosomal protein, S1, that is more loosely bound than the others. The result was we had a very pure sample of 30S subunits.

Before binding IF3 to the 30S subunits, we decided to check if these 30S subunits were good enough to crystallize at all using what was already known. Even though the best reported crystals to date were not good enough to yield an atomic structure, even a low-resolution structure could still tell us a lot about how the RNA was folded. We'd also be able to place proteins whose structures had already been solved in isolation and slowly build up a piecemeal model of the subunit. That could keep us busy until we got better crystals. Within a couple of months, we had reproduced the original Russian crystals. They were pretty small, and some initial tests showed they diffracted to quite low resolution, worse than the best ones reported so far. However, they were slowly getting bigger.

Just as we had embarked on the 30S project, Steve White and I were invited to a meeting in Sweden on the structural aspects of protein synthesis, that is, the structures of anything to do with ribosomes. The organizer, Anders Liljas, was collaborating with Maria Garber on individual ribosomal protein structures, and they were friendly competitors of our effort with Steve. Even though I was already bored with the work, it was an opportunity to go and see what was happening. Not only would Ada be there but also Peter, so I would see what the competition was up to. There was also another reason go to: I could visit the LMB in England again, since it was practically on the way to Sweden. This was not merely for a nostalgic social visit – rather, I had come to the conclusion that it was the best place for a risky project like the 30S structure and wanted to see if I could work there.

Ever since embarking on the 30S project in Utah, I was excited but also quite fearful. What if it took me years to get good crystals? Even if I got them, what if my ideas of how to solve the structure didn't work out? The problem with doing it in a university setting like Utah was that my research depended on grants, which last only a few years before you have to renew them.

These grants are typically funded by the National Institutes of Health, after being reviewed by a panel of a dozen or more experts in your field. In theory, that is a great way to fund science, and it has worked extremely well. But one scientist described the system as being like your favourite restaurant – you don't want to go into the kitchen to see how the food is being made. There are two problems with these expert panels. On the whole they tend to be too conservative and lack the vision or confidence of judgement to support bold, original proposals, preferring incremental ones where the work is clearly feasible. The only way to ensure you fill the panels with the really top scientists who are unafraid to back adventurous ideas is to make it like jury service, so if you receive an NIH grant, you are required to serve if called. The other problem is that each panel receives over a hundred proposals, each with more than fifty pages of dense information. In practice, each application is read in detail by only a couple of people (called the primary and secondary reviewers). Because only a small fraction get funded, even if one of these reviewers isn't enthusiastic about a proposal, it is essentially doomed. Once when I had argued against the primary and secondary reviewers to try to rescue a proposal, everybody else just split the difference and gave it an average score, so it was doomed anyway. So although the process is fair on paper, in practice it can be somewhat arbitrary, particularly when it comes to risky or highly original proposals.

Having served on these panels, I could just imagine how they would receive a proposal from me on the ribosome. I didn't have crystals but *thought I had an idea* of how to get good ones. And although a well-funded group in Germany had been unable to solve a structure with good crystals for years, I just *had an idea* of how to go about it. I could just hear the peals of laughter that would go around the committee room as they dumped my proposal into the bin. The other option was to divert resources from my existing grants to do the work on the side, which many scientists have done to do their most creative work. But given the competition and the need to focus on it completely, I didn't think this was a viable strategy. Also, if the ideas didn't work, I would lose my grants and would have to start from scratch, and I might never recover. Every research university has people who lost their research funding and ended up being second-class citizens, whom their department tries to force out or marginalize.

But the LMB was somewhat different. They understood that some problems can take a long time and, more importantly, had a lot of people who knew what it took to crack them. So, even though I'd already approached them about a job when I had just returned from my sabbatical, I wrote again to Richard Henderson, who by this time had become the director of the LMB. This time I had a concrete proposition for them. I had ideas for how to crystallize the 30S subunit and solve its structure and was interested in moving to the LMB to work on it. Could I stop off in Cambridge and discuss it with them on my way to Sweden? Richard said he'd be happy to see me.

The visit to Cambridge was unlike any 'job interview' I've ever had. For starters, there was no job. For another, they never once discussed any of the usual things like space, equipment, or, God forbid, salary. Rather, I gave a talk on the structures we'd solved of various ribosomal proteins. Then all afternoon, I had a completely frank discussion about the ribosome problem with Richard and Tony Crowther, who, like Richard, was a renowned electron microscopist and had become a joint head (or co-chair) of the Structural Studies Division. We discussed who was doing what, why the field was stuck, how I might tackle it, what sort of difficulties I might encounter, and how long I thought it might take to get to an intermediate step where we could see recognizable features in the maps of the 30S subunit. This sort of free-flowing intellectual give and take was quite unusual for a job interview, even if there wasn't actually a job. 'Let's keep in touch,' was their verdict. Despite the non-committal response, talking to them had made me feel it wasn't such a harebrained idea after all, so I went from there to Sweden feeling pretty enthusiastic.

Many meetings are held in remote resorts to force people to discuss science instead of going off shopping or seeing the sights. Tällberg, a quaint village on the banks of Lake Siljan in Dalarna province many hours north of Stockholm, fitted the bill perfectly. Anders Liljas had started organizing meetings in Tällberg because he had grown up there himself and knew a charming resort hotel that was suitable for holding conferences of about a hundred or so people. He was a big jolly man with a cosmopolitan outlook, but he was still rooted in his traditional rural Swedish upbringing (after he retired, he went back to his ancestral home in Tällberg). He was known by just about everyone in the ribosome field, and his strength was that he liked to think the best of people and wanted them all to get along. Although this made him a popular and trusted figure, his diplomatic skills were to be sorely tested in the next decade.

For the first time, I was going to hear Peter tell us what the Yale group had been doing with their crystals. What sort of data did they now have and how had they gone about obtaining the phases that were essential to determining a structure? With normal proteins, a single heavy atom like mercury or gold can give enough of a signal to be seen directly in the Patterson maps we discussed earlier, which have peaks that correspond to the distances between the different heavy atoms in the repeating unit of the crystal. But as the molecule of interest becomes larger, the relative signal from the heavy atom becomes smaller compared to the rest of the protein until it becomes practically undetectable. So for larger molecules, people had tried using heavy-atom clusters to get started. These clusters were small inorganic molecules that contained several atoms close together, typically tantalum or tungsten. At low resolution, the many atoms in the cluster would act like a single 'superheavy' atom, so the signal from them could be quite large. Just such clusters had been used before, for example to solve the nucleosome core particle or a large enzyme called rubisco, which plays a key role in carbon fixation by plants. But the 50S was much larger than any of these. In looking into this problem, Ada had thought of using large tungsten clusters. These clusters had up to thirty atoms and might do the trick.

Still, it was not clear that these large clusters would bind to the ribosome at all, or even if they did, whether they would give a clear signal. In Ada's talks in Victoria or Seattle in the previous two years, there was no clear evidence that a heavy-atom cluster had bound to the ribosomal subunits in her crystal. But then, in Peter's talk in Tällberg, there it was: a giant spherical blob in their Patterson maps. It was the first direct proof that they could see a heavy atom bound to the ribosome.

They had gone on to confirm the result by an independent strategy. Apparently, Peter had been impressed enough in Victoria by Joachim Frank's electron microscopy maps of the ribosome that he thought they might be a useful starting point to analyse data from the crystals of the 50S. The method is called molecular replacement, and the principle is this: with X-ray data, as we discussed earlier, you measure the intensity of the spots but don't know the phase, and you need both to reconstruct a three-dimensional image of the molecule. Normally, you would use the heavy-atom method that Max Perutz and his colleagues had invented to obtain the phases. But if you can find a similar molecule whose structure is known, then you can figure out how this 'test' molecule would pack in your crystal and calculate its diffraction pattern. You'd then use the phases from this test molecule with the actual measured intensities from the data to get an approximate starting structure for your molecule, then calculate phases from that starting structure to get better phases, and gradually bootstrap your way to a correct structure. People like Steve Harrison had previously used three-dimensional images from electron microscopy to get started on some of their virus structures, so Peter reasonably wondered whether it would work with Joachim's maps of the ribosome.

At the meeting, Peter showed how they had been able to use one of Joachim's maps to figure out how the 50S subunit would sit in their crystals. The approximate phases from this exercise predicted the peak they had already seen in their Patterson maps. The resulting map of the 50S from their X-ray data was still quite low resolution but looked plausible and recognizable. There was no longer any doubt that their strategy was working.

Then Ada spoke. She started off by showing data on the 50S subunit, which showed that the cell dimensions, which is the distance between repeating molecules in the crystal, changed as you collected data. This meant that because of radiation damage, the crystals were changing while you were collecting data on them; she said they were unusable, and the implication was that the Yale group was wasting its time. The irony of Ada rejecting the crystals her own group had developed, while the Yale group was happy to work with them, was not lost on me.

I was a bystander in this rivalry over the 50S subunit, so up to this point, I was relaxed. That is, until Ada said she now had managed to get the 30S subunit to diffract well when bound to a heavy-atom cluster. Presumably, she had been looking for a heavy-atom derivative, and one of them stabilized the 30S subunits, enabling them to become better ordered in the crystal. She then showed a diffraction pattern showing lots of spots. I couldn't believe what I was seeing and hearing. What I thought would be a quiet niche for

Figure 9.2 At the Tällberg meeting: Ada Yonath, Steve White, Francois Franceschi, and the author *(courtesy of Isao Tanaka)*

me would now mean a head-to-head competition with Ada, and we weren't even sure we could get good crystals yet. I was in turmoil for the rest of the meeting. During a break, Steve White and I went for a walk in the nearby woods around the lake with Ada and Francois Franceschi, along with Isao Tanaka, a crystallographer from Japan, and I had to put on a good face the whole time.

I mulled over the situation on what seemed like an interminably long plane ride back to Utah. I briefly considered just giving up, but then realized that Ada had obtained good crystals of the 50S subunit many years earlier, and she still hadn't solved it. Just because she now had good crystals of the 30S subunit didn't mean she'd suddenly figure out how to solve it either. The Yale group had shown that the heavy-atom clusters were fine to get low-resolution maps, but as you get to high resolution, the heavy atoms in the cluster don't behave like a single superheavy atom. At finer detail, they look like a lot of separate atoms, and their signal drops. So my idea of using special atoms with synchrotrons to solve the structure using anomalous scattering still seemed to be the only way forward to get a high-resolution structure that would yield an atomic model of the ribosome.

In the end, what really swayed me was the realization that the structure of the ribosome was the most important goal in my field. There seemed a narrow window of opportunity, and since I had a clear idea of how to attack the problem, it would be a mistake to be dissuaded by this new development. Furthermore, even if I wasn't one of the first groups to solve an atomic structure, the ribosome was such a complicated machine that understanding it would take lots of structures in different states of the process. There would be plenty of interesting work for years to come, and to be best positioned for that, I should get started sooner rather than later. Anyway, there was no time to lose: instead of Yale, it was I who would be in direct competition with Ada's large, well-funded group.

I felt I had to tell Richard of the new situation. Fortunately, he was not in the least bothered by it. He felt I had at least as good a shot as anyone and thought the fact that I had solved many different structures was good experience for tackling a hard problem. But there was still no job offer. He said they were interested in my coming to the LMB to work on the ribosome, but they didn't have any space at the moment. However, they were expecting to acquire some, and when they did, they would get in touch.

Given Ada's improved crystals, we could not afford to wait for the LMB. I told my lab about the meeting in Tällberg, and it was full steam ahead. We had to adopt a two-pronged strategy. The first was to continue to see where the crystals of the 30S would lead us. Perhaps we too might be able to stabilize them in some way, as Ada seemed to have done. The other was to try to lock them with IF3 and try to get crystals of that, which would also be more interesting because it would show the 30S in the act of doing something like binding a protein that is required to initiate translation. Fortunately, this idea didn't need to work, or I'd still be waiting.

Figure 9.3 **Malcolm Capel** *(courtesy of Malcolm Capel)* **and Bob Sweet** *(courtesy of Brookhaven National Laboratory)*

Our immediate task was to see how good the crystals we already had were. When I went to Brookhaven to collect data on a different project on Bob Sweet's beamline, I also took several frozen crystals of the 30S subunit. We decided to look at them on the adjacent beamline, which was more intense and would give us a better idea of their quality. It was run by my old friend and colleague Malcolm Capel. Malcolm was a big guy with a large bushy beard and long hair that he wore in a ponytail. He had grown up in a Mormon family in Utah, but despite this fairly conservative upbringing, he was a highly irreverent and irreligious character who liked his beer and partying and peppering his language with swear words in new and creative ways. As the evening wore on at parties and the beer started to work, he would become incredibly witty, making hilarious and very frank remarks about everyone we knew, including those present. Because we had worked on the same neutron project at Yale (although some years apart) and we shared a connection through his wife, Sue Ellen Gerchman, who was my technician at Brookhaven, we had gone on to become good friends.

I knew that Malcolm had an interest in ribosomes because of his background, and as an old friend and colleague, I implicitly trusted him. So I had told him about our project, and he said he'd help us in any way he could with his beamline. On my visit, we did test shots of lots of crystals on his beamline, and although they were a promising start, the spots did not extend out to high resolution.

This was particularly disheartening because just a little while earlier I had encountered Nenad Ban, who had been collecting data on Bob Sweet's beamline before me. Nenad was an obviously intelligent and good-looking Croatian with a boyish face and engaging smile. He had done his PhD with Alex McPherson in California before joining Tom Steitz's lab as a postdoc. When he found out during his interview that Tom wanted to tackle the ribosome, Nenad was instantly excited because he had long been interested in it – he sometimes shows a drawing of the ribosome he made when he was still a schoolboy. Nenad told me that his biggest fear was not the challenges of working on this daunting problem but rather that Tom might give away the project to someone else before he actually showed up some months later.

When I arrived at Bob's beamline, the last diffraction image that Nenad had collected on their 50S crystals was still up on the screen, and my heart sank. Unlike our weak diffraction, it showed strong

Figure 9.4 The Yale team working on the 50S subunit: Nenad Ban, Tom Steitz, Peter Moore, and Poul Nissen *(courtesy of Tom Steitz)*

diffraction spots all the way to the edge of the detector, even on a relatively low intensity beamline. Nenad's confidence was also intimidating: he was in a hurry to catch the ferry back to New Haven and asked if I would take out the tape that was backing up his data when he was finished. He didn't seem the slightest bit worried that I could take a sneak look at it.

Soon afterwards, we learned that the Yale group had crossed the 10 Å resolution threshold. They eventually published a paper in *Cell* in which their maps showed the right-handed double-helical twist of segments of the RNA. They rather pointedly mentioned that their map showed 'density features expected from other structural studies' and went on to say that the work 'constitutes an important beachhead for launching an attack on the structure of the ribosome at higher resolution.'

I was already worried about Peter and Tom being formidable rivals. Nenad, whom I had met, was obviously first rate. I noticed that the author list included Poul Nissen, another top-notch young crystallographer who had joined them from Århus in Denmark after solving a very important and tricky structure on the complex that delivers amino acids to the ribosome, consisting of tRNA with its attached amino acid and a protein factor called EF-Tu. So Yale had assembled a crack team, while we didn't even have good crystals yet. Hopefully they wouldn't figure out my idea of how to get to high resolution too quickly. We were falling further behind, but there was nothing to do but push on.

To test our idea of locking the 30S subunit down by forming a complex with IF3, I contacted Joachim Frank about collaborating on doing a structure of it by electron microscopy. This would be a low-resolution structure, but since very little was known about how IF3 bound at all, it would be interesting in itself. It would also give us some feedback about whether the idea of locking the 30S was working. Importantly, now that the Yale group had used an electron microscopy map to get started with phasing their X-ray data, it might help us do the same with the 30S. At the time, Joachim wasn't aware that this was really a prelude to our attempting to crystallize the complex, but since just knowing roughly how IF3 bound to the 30S was interesting enough, he said yes, and had his postdoc, Raj Agrawal, collaborate with John McCutcheon on it. John soon made the appropriate complex and went off to Albany to work with Raj on doing an EM structure of IF3 complex.

In the meantime, our crystals had grown larger. On the next trip to Brookhaven, they seemed to do a lot better on Malcolm's beamline. The best of them seemed to go to about 5 Å resolution – close but not quite there. Bob Sweet came by and said they had a new high-intensity beamline and asked if I'd like to try my crystals on it. I had nothing to lose, so I stuck a crystal on and nearly had a stroke. The crystal diffracted to beyond 4 Å resolution, with diffraction spots going out almost to the edge of the detector! The spots at high angles were quite weak, which is why we had needed a very intense beam to see them. But it meant the crystals were good after all, even without binding IF3, as I'd originally thought would be necessary, or stabilizing them with some (yet unknown to us) heavy-atom cluster, as Ada had done. Perhaps all it had taken was to make sure the preparation was very pure and homogeneous, with the variable S1 protein removed. And possibly growing them very slowly over a period of weeks in the cold room at 4 degrees C also helped produce better crystals. Whatever the reason, to our pleasant surprise, we had caught up with the other groups with our well-diffracting crystals. Now we had to figure out what to do with them.

CHAPTER 10

A Return to Mecca

SOME TIME AFTER RETURNING FROM Sweden, I got a phone call from Ada out of the blue. I had described our structure of S15 in Tällberg, and Ada asked if she could mention the results in a review she was writing. I thought this was an odd reason to call me. It could easily have been dealt with in an email. I told her that the structure was due to be published soon, and she could certainly cite it. Then, somewhat mischievously, I said, 'Well, Ada, it was nice talking to you.'

'There is one other thing,' she immediately replied, coming to the real reason for her phone call. She said she had heard we were working on the 30S subunit. I didn't want to have to discuss with her where we were and what we were doing, but I didn't want to lie outright either. So in a Clintonian manner (Bill, not Hillary), I said, 'We're thinking about it.' After all, we *were* thinking about it: night and day. Ada added that they'd made great progress, and in her maps she could now recognize the proteins whose structures we had already solved in isolation. If we had new ideas, she'd be happy to consider them. I politely told her that for now we wanted to give it a shot ourselves to see how far we could get with our own ideas. I didn't know this at the time, but Peter Moore recently told me that she had paid a surprise visit to Yale around then and offered to collaborate with them too.

Collaborations work best when the people involved are good friends, enjoy working together, and have complete confidence in each other, or when they have complementary expertise that allows them to tackle something neither could do alone. They also require a willingness to give up complete control of the project and to share credit in ways that might not seem fair to all parties. At this point, neither I nor Yale were interested in collaborating with anyone else.

Ada's call left me in a complete panic. Clearly, I had not been as discreet as I had thought. And what did she mean she could recognize our protein structures in her maps? If that was true, she was so far ahead of us that perhaps we shouldn't even bother. Distraught, I went down the hall to see my colleague Chris Hill, who just laughed. He thought it was pretty unlikely that the first thing someone would do after obtaining great maps would be to call a potential competitor and tell him about them. I felt better, but the whole thing stressed me out because it brought home the reality of the competition that I had originally tried to avoid.

Finally, Richard Henderson wrote. The space had come through, and they were interested in proceeding with hiring me. I suddenly had to make what was one of the hardest decisions of my life: whether to gamble everything on going to the LMB and work exclusively on this project. If I stayed in Utah, I'd have to hedge my bets by working on other safer projects simultaneously. But these projects would slow me down on the 30S, and now that the race was heating up, I felt I needed to focus on it completely. There was a narrow window of opportunity, and if I didn't grab it, I would watch other groups do it and regret it for the rest of my life.

Vera and I enjoyed living in Utah, and many of my colleagues there had become good friends. For a while, I was torn. Then I decided to seek some advice from two people whom I particularly respected because they had tackled similar hard problems. The first was Peter Moore. I didn't tell him about the 30S project but asked him whether he thought moving to the LMB was a good idea. He said Utah was a fine place, but the LMB was unique, and if I had a chance to go there, I should seriously consider it.

The second was Steve Harrison from Harvard, who was one of the foremost structural biologists of his generation. He had burst onto the scene when quite young, tackling what seemed an impossible problem at the time, the structure of an entire virus, when computers were primitive and there were no synchrotrons. He also showed courage in his personal life, being unabashedly gay at a time when most people were afraid to come out of the closet. I doubt he ever imagined that one day he would be legally married to his longtime partner, Tommy Kirchhausen, a well-known cell biologist at Harvard.

Steve was also famous for being brutally frank. He publicly characterized crystallization in space – an expensive effort to send protein solutions up to the space station to try to get better crystals under zero gravity – as a waste of time and money. He once began an advisory meeting at Brookhaven by roundly chastising the director of the synchrotron for presuming to tell biologists how to do their experiments. I felt that if there was anyone who would tell me bluntly that moving to the LMB to work on the ribosome was a bad idea, it was Steve. As it turned out, he didn't. Instead, he thought that the field would benefit from competition, and now that others like Tom and Peter had entered the race, there was no reason why I shouldn't. I should consider what the LMB could provide that Utah couldn't.

I didn't know it at the time, but Steve is a good friend of Ada's, and it says something about his integrity and objectivity that he was so encouraging. Over the years, he went on to become a close friend of mine too. Because of his love of chamber music, he has even got to know and on occasion supported my son Raman, now a cellist.

In the end, after a long walk with Vera, we made our decision. We would leave our families (including our grown children) in America and go to England, where I would take a large salary cut to stake all on the ribosome at the LMB. Vera had uncomplainingly followed me around the US throughout my career, even when I went across the country to go back to graduate school right after Raman was born. I was lucky that she was a self-employed artist and writer, but even so, each time she had to uproot her life and move away from her home and friends. She agreed to do it once more, if I felt it was important for my work. However, she said, it would be the last time. So far, she has kept her word.

Dana Carroll, who had hired me just a few years earlier, was visibly shaken to hear the news. He made the prospective salary cut even more painful by immediately giving me a raise, but deep down he knew it was not going to matter. In the end, he, Wes, and Chris were wonderfully supportive even after I had told them I was leaving. Rather than being angry, I got the sense they were rooting for me.

In any case, since I had committed to going to Cambridge to work on the ribosome, we needed to make as much progress as we could to make up for any time lost to the move. Brian Wimberly, who had initially been cautious about working on a very risky project for his second postdoc, had now solved two structures, including an important one of a ribosomal protein bound to a piece of RNA. At this point, his postdoc could be considered a success, and he, too, wholeheartedly threw himself into the 30S project.

We now had to figure out how to obtain all the compounds we might need to produce an initial structure from the crystals. The first thing I did was go through the periodic table and get salts of every metal that would provide a strong anomalous signal at X-ray wavelengths that synchrotrons could produce. These were mostly the lanthanides like holmium, ytterbium, europium, and so on. To use these elements to get phases and thus a map, you had to first locate where they were bound in the crystal. There were computational methods that should allow you to locate them directly, but they had never been tried on a problem like this before. What if the signal from each of them was too weak?

However, the Yale group had already shown that at least one heavy-atom cluster could be seen directly in Patterson maps. Even though you would only get low-resolution phases from it, you could use those phases to locate where the other atoms like the lanthanides were. We were in a race now and couldn't afford to try one thing at a time. Instead of first trying to see if directly locating the lanthanides would work, I felt I had to cover my bases. So like Ada, and later Tom and Peter, I too wrote to Michael Pope, an inorganic chemist at Georgetown University, and asked him if he could provide me some of his tungsten cluster compounds. He was very friendly and provided me with his entire series of compounds, which were inorganic molecules containing anywhere from eleven to a staggering thirty tungsten atoms. He may well have wondered at the sudden interest in his tungsten compounds from various crystallographers he'd probably never heard of. I was able to get tantalum bromide from Gunther Schneider in Stockholm, who had used them to solve a large protein complex. Science does indeed depend on the kindness of strangers.

The strategy was clear. I could get low-resolution phases from one of these clusters (or from an electron microscopy map) as the Yale group had done. Then, if I used those to locate where the atoms with strong anomalous signals bound in the crystal, I'd be able to bootstrap my way to high-resolution phases and be off to the races. After all, my calculations had already shown that the signal ought to be strong enough. But these calculations assumed the kind of diffraction you'd get from a typical protein. The diffraction from the 30S was much weaker, so I realized I couldn't be sure it would work. You'd need to do the experiment and just hope that the data were good enough.

I saw a better alternative when, around this time, a paper came out on the structure of the largest piece of RNA determined to date. It was part of the RNA molecule that Tom Cech had shown could cleave itself without the help of a protein enzyme, for which he had shared the Nobel Prize. Cech's postdoc, Jennifer Doudna, had

Figure 10.1 Jennifer Doudna and Jamie Cate *(courtesy of Jamie Cate)*

started the project to determine its structure in his lab in Boulder, but then continued it at Yale when she joined its faculty.

Jennifer was a brilliant scientist who had done her PhD with Jack Szostak at Harvard, where she worked on RNA catalysis. Following her postdoc with Cech, she went to Yale, where she had a remarkable series of successes and continued an illustrious career in Berkeley. In the last few years, she has taken the world by storm for her part in developing the CRISPR-Cas method of modifying genes, for which she has won numerous awards and probably will win the Nobel Prize. Unusually for a scientist, she was also once featured in *Vogue*. The graduate student who began working on the project with her in Colorado, Jamie Cate, followed her to Yale. He did more than just follow her. They soon entered into a relationship and are now a star couple like Tom and Joan.

The paper by Jamie and Jennifer had used the same MAD method I was thinking of for the 30S subunit, but in addition to the lanthanide metals that I was considering, they had hit on an even more magical bullet. It was a compound called osmium hexammine that nobody had used before for phasing crystal structures of large molecules. From the way it bound to their RNA molecule, I figured it would bind to the ribosome in dozens of places, unlike lanthanides, which might only bind in ten or twenty sites at most, so it was likely to give a much better signal.

Alas, it was not commercially available. The paper said they had obtained the compound from Henry Taube, a well-known inorganic chemist at Stanford. I wrote to Taube, but here my luck almost ran out. He wrote me a terse and somewhat irate letter saying he had run out of the material, and since his grant was terminated, he didn't have the resources to make any more. I, a theoretical physicist turned biologist, certainly didn't have the skills to make it myself. Fortunately, science depends on the kindness not only of strangers but also of friends.

The friend in this case was Bruce Brunschwig. Vera and I had got to know Bruce and his wife Karen over the dozen years at Brookhaven. From the moment I first met them soon after my arrival there, they struck me as people I'd want to know better. Bruce, who had grown up in Philadelphia, had an irreverent East Coast Jewish sense of humour that I particularly enjoyed. We became (and remain) very close friends partly because we hit it off and partly because our children grew up together and went to the same school. They lived a couple of blocks away, and we would often spend weekends together on outings where the main feature would be the long naps that Bruce and I would take after lunch.

So in desperation, I asked Bruce, a trained inorganic chemist, if he could make me some osmium hexammine. He took a look at the synthesis and said he probably could make it for me in a couple of weeks with the help of his technician. It is typical of his character that he declined to be a co-author on the papers that would make me well known, saying it was just a routine synthesis. Others have been authors for far less, and his help made the difference between success and failure.

So finally, I had assembled the compounds that I thought we'd need to use with our crystals to get a structure. The collaboration with Joachim and Raj on the IF3 project was also proceeding well, and we would soon have electron microscopy maps of the 30S as well to perhaps use to get initial phases. With a steady supply of crystals being produced slowly in the cold room, we were ready to blast off.

While all this was happening, I got yet another shock. One of my colleagues in Utah was Brenda Bass, who had made important discoveries about how mRNA was modified (or 'edited'). Soon after I had arrived in Utah, I found out she was in a relationship with Harry Noller, the biochemist who was well known for his work on ribosomal RNA. I remember being quite surprised, at least in part because Santa Cruz was not exactly next door, and thinking Harry was rather lucky to persuade someone like Brenda to enter into a long-distance relationship. However, I thought no more about it until, while attending a picnic at Brenda's place, I encountered Harry, who was visiting from Santa Cruz. We struck up a conversation, and he told me they were working on crystals of the ribosome. Crystals! How could Harry, a biochemist with no experience in crystallography, take on the entire ribosome? It turned out that he had hired Marat Yusupov and his wife, Gulnara, who had been part of the team in Russia that had produced the first crystals of the whole ribosome from *Thermus*. After their collaboration in Strasbourg had ended, they had been frustrated at being shut out of ribosome crystallography.

So when Harry published a paper showing that he could reassemble the head of the 30S subunit from individual proteins and a piece of RNA, Marat wrote to Harry asking if he could come and work on the structure of this piece of the 30S subunit. Harry was not a guy to do things by halves (or a third of a third of the whole ribosome in this case). The head of the 30S would tell you almost nothing about how the ribosome worked. Why not come and work on the entire ribosome?

Marat and Gulnara were thrilled, and the couple went off to Santa Cruz to start working on the entire ribosome. Marat told me he suggested to Harry that they collaborate with someone well known who had experience in solving hard structures. Instead, preferring to keep it in-house, Harry hired a young ambitious postdoc – none other than Jamie Cate. Although I was in denial at first, I realized that Jamie, who was no fool, was one of the few people who would almost certainly see the exact same path to phasing data from ribosomes that I thought was my secret idea.

With the Yusupovs producing crystals using their years of expertise starting in Russia, and Jamie's experience at solving a large RNA structure using exactly the method that would be useful for the ribosome, Harry had assembled the perfect team. I wasn't at all happy that, in less than a year, what I thought was a niche for myself had turned into a four-way race. For the rest of my stay in Utah, I could not help being paranoid about discussing the ribosome project with Brenda. This was a pity because she was someone I admired and with whom I shared a lot of scientific interests.

Soon after the decision to move to Cambridge, John McCutcheon came to see me with an embarrassed look. After initially being very enthusiastic about the move, he said he couldn't go after all. He was now in a relationship with a fellow grad student, and it would be too complicated. The technician, Joanna May, also had a family and was not moving. Suddenly, my already small group was going to be half its size as a result of moving.

I went to the annual students' day at the LMB in the hope of persuading a grad student or two to join the effort. One of them, a well-qualified German, listened to my pitch and said, 'There's a large group in Germany that has been trying to do this for twenty years! What makes you think you'll succeed?' I tried to tell him that even if we didn't succeed right away and got scooped on the first structures, there would still be plenty of work left for us to understand the ribosome. Almost right after our interview, he signed on with Kiyoshi Nagai, my friend and future lab neighbour. A couple of other students, including one from Cambridge, seemed to be completely clueless about the problem. I thought they'd only slow us down.

Despondent, I spent the flight home wondering whether I was shooting myself in the foot by moving. Shortly after my return, I wrote to Tony Crowther at the LMB saying that I was rethinking the whole move. Fortunately, in the next few weeks, two people courageously agreed to join my lab at the LMB without ever having met me: Andrew Carter, who joined me from Oxford as a PhD student, and Ditlev Brodersen, a postdoc who came highly recommended to me from Århus in Denmark, where I knew his supervisor, Morten Kjeldgaard. As I was to find out, I could not have had two people who complemented Bil and Brian more perfectly. Their joining me was a stroke of luck and meant we would have a viable team at the LMB.

Moving to a new lab always costs time, and moving to a different continent in the middle of a race seemed insane even to me. So I wanted to make sure that we gave ourselves a good head start. We had electron microscopy maps of the 30S subunit from Joachim as a result of the IF3 project. However, I failed to locate the 30S in our crystal using these maps with the X-ray data – perhaps the maps were at too low a resolution and perhaps the 30S was too thin and flat an object. But since the Yale group had shown you could directly see heavy-atom clusters in a Patterson map, trying to use electron microscopy to get started wasn't actually necessary, even to get low-resolution phases.

So Bil ended up soaking dozens of crystals in each of the many compounds I'd accumulated, and with Brian, in early March 1999, we were off to the synchrotron at Brookhaven to see what we could get with them. Unlike Ada's unusual observation of a heavy-atom cluster actually improving diffraction, our best diffraction was still from the 'native' crystal to which we hadn't bound anything. As is much more typical, the crystals soaked in the various compounds all diffracted worse to varying degrees.

We had chosen the wavelength of the X-rays to maximize the anomalous scattering from the heavy atoms and thus increase the small differences in the intensities of the symmetry-related spots. Those differences contain the signal from the heavy atoms, and a Patterson map using the differences would produce a peak showing where the heavy atoms were. As the data were being collected, we processed them on the fly and calculated one Patterson map after another.

Nothing seemed to be working, and I was slowly getting exhausted and depressed. Then, after midnight, a Patterson map from a crystal soaked with a seventeen-atom tungsten cluster showed two large unmistakable peaks. Brian and I looked at each other and instantly leaped out of our chairs and did a high five with a huge shout, startling a lone physicist who happened to be nearby. We quickly did some additional experiments on the same crystal to confirm the peaks, and when we saw them again, there was no doubt at all.

We returned to Utah euphoric. Soon afterwards, a low-resolution map calculated from just the seventeen-atom cluster clearly showed an outline of the 30S subunit and how it was packed in the crystal. We had our foot in the door.

Maddeningly, only a month after having made this exciting breakthrough, it was time for me to leave Utah. Bil and Brian were to continue working there with Joanna for a few months, help close up the lab, and then follow me to England. Having sold our house, Vera and I took a flight for London on April 15, 1999.

CHAPTER 11

Coming Out of the Closet

IT'S ONE THING TO VISIT a foreign country for a year and quite another to burn your bridges and move there. All countries have their peculiarities, but we become used to our own, whereas those of other countries strike us as strange. So in America, I had become used to the idea that lots of otherwise reasonable people own guns for no good reason, public transport is virtually non-existent in most places, and many people live in suburban sprawls and drive everywhere. During our sabbatical in England, we had noticed things like rigid bureaucratic rules administered with an earnest smugness, queues for almost anything forming in ways that were impenetrable to foreigners so you were never actually in line, customer service that was an oxymoron, or – my favourite – locals feeling that 'we've always done it that way' was a perfectly reasonable response when questioned about some completely silly practice. But things that had seemed charming and quaint if a little odd during our sabbatical were annoying when they became a permanent part of our lives.

We had sold our five-bedroom house overlooking the Salt Lake valley and the Wasatch Mountain Range and were renting a place owned by the MRC. From the very first week, we started looking to buy a house. Before committing to the job, I had enquired what a modest terraced house would cost in Cambridge and figured we could just about afford one. Between the decision and the move, prices had risen by almost 50 per cent and were continuing to rise almost weekly. For a while, we were outbid on virtually every house, and our inability to buy anything at all added to Vera's unhappiness about leaving her friends and nice home in Utah. It probably didn't help that right from the start, I had a near obsessive focus on the 30S structure.

Normally a move would have slowed us down considerably, but for a couple of reasons we may actually have speeded up our progress instead. Since we had collected the data we needed for this phase of the work, I could not have moved at a better time because the LMB had excellent computing facilities. This meant that I could speed up the work by trying lots of calculations in parallel, see what worked, and then use that as a guide to do the next set of calculations.

It turned out that virtually every compound Bil had soaked into the 30S crystals was useful. The seventeen-atom tungsten cluster was the only one whose signal was large enough to see directly as peaks in sections of the Patterson maps. But although it wasn't obvious from a direct inspection of the Patterson maps, all the other compounds had also bound to the 30S subunit.

When the signal is weaker, there are programs that can find the heavy atoms in a nearly automatic way. One of them was a program called SOLVE, written by Tom Terwilliger from Los Alamos. Tom is one of a small group of brilliant computational crystallographers – the people who write the software that everyone uses. He is a cheerful guy with a great sense of humour who works at Los Alamos National Lab where his wife is a ranger in the Santa Fe National Forest nearby. I had used his program when I'd done a comparison of various ways to solve MAD structures. His program was as good as any of the others and was by far the easiest to use. He had written SOLVE to try to automate the problem of not only finding the heavy-atom derivatives, but also using them to calculate phases and produce the electron density maps – the three-dimensional images.

Brian started using SOLVE with our 30S data, and by the time I had settled into the lab in Cambridge, he found that SOLVE had identified peaks for heavy-atom positions in every compound Bil had soaked into the 30S crystals. This included all the other clusters and the various lanthanides and osmium hexammine.

Initially, the program would identify only the strongest peaks, but by combining the data, it would slowly bring out some of the weaker ones. Combining them wasn't so straightforward because – as is often the case – adding a heavy-atom compound would change the crystals themselves. They were no longer isomorphous, which is to say the 30S subunits in the crystals had themselves changed slightly as a result of binding the compound. So it was not possible to combine the data from different crystals easily. We had about fifteen or twenty data sets and we needed to figure out the combination that would give us the best maps. Being at the LMB, with its multiple computers nodes, was especially helpful because I could now try lots of different combinations in parallel and compare the various maps.

Another odd advantage of my move was the time difference between Cambridge and Utah. I would set up a number of computing jobs and, at the end of each day, I would email the results to Utah, which was seven hours behind. Brian and Bil would look at the maps during *their* day and tell me what things were working and what weren't. So when I came in the next morning, I'd have my feedback. This effectively extended our working day by about seven hours, so we were working almost around the clock.

Slowly, the molecule appeared before our eyes. Initially, we could see its broad outlines – where the molecule was and where it touched neighbouring molecules in the crystal lattice. Then its detailed shape began to appear. As the maps improved, we were able to find even weaker sites – ones that SOLVE couldn't identify automatically – and throwing them into the calculations made the maps even better.

Suddenly, a month after I had arrived in Cambridge, there it was: a long RNA double helix that ran right down the face of the

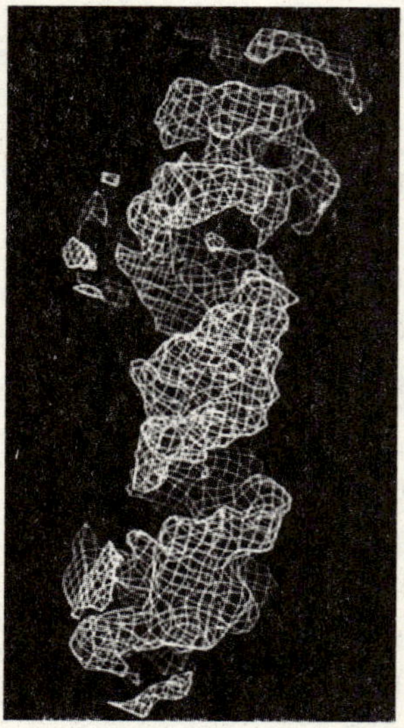

Figure 11.1 An exciting moment – being able to see a clear RNA double helix, with little bumps for the phosphate groups on each strand

30S subunit. Unusually for me, it was fairly late, and I rushed out of the graphics room in the LMB and found Richard Henderson, who was famous for working late into the night. He agreed it looked like a double helix. Excitedly, I sent it off to Utah. I almost wished I could have been there when they saw it.

Brian very quickly identified the outlines of the 30S and its characteristic shape in the map. He then went on to find lots of other double helical regions of RNA. We knew that the RNA in the 30S formed around forty of these helices, although some of them were quite short, unlike the long helix (number 44 or h44) that we'd first seen. RNA helices are A form, which is named after the dehydrated form of DNA that Rosalind Franklin first observed, unlike the more typical B form of normal DNA that she saw when DNA was hydrated. We could easily see the narrow but deep major groove and the wide but shallow minor groove that characterize these A-form helices. Our best maps were at around 5.5 Å resolution, which meant we could even see phosphate groups as a series of bumps along a ridge as we followed the helical turn of RNA. Our strategy was working beyond our expectations.

At this point, my colleague Daniela Rhodes suggested that we should report our results in *Nature.* Daniela is well known for her work on chromatin, but she was also the person who over two decades earlier had done important work on tRNA with Aaron Klug and Brian Clark. We had become good friends during my sabbatical, and she had been strongly supportive of my move to Cambridge. She told an editor at *Nature* about our results, and he got in touch to say he was very interested in publishing them. We thought a short report describing our progress would be a good way of staking our turf. These short reports in *Nature* are called letters, for historical reasons, as opposed to articles, which are longer and more substantial papers. But longer is not always better or more important. One of the most famous papers in *Nature*, the Watson and Crick description of the double helical structure of DNA, is an amazingly short letter of only about eight hundred words.

But something seemed wrong with our maps. We could see a lot of RNA in them, but there seemed to be no sign of proteins. They had to be there since there were about twenty of them in the 30S subunit. Maybe because they are not as dense as the RNA, they weren't showing up in our maps. I was puzzling over this when I noticed that some of the density looked like tubes that were a lot thinner than the RNA double helices. In fact, they seemed the right size to be the alpha helices that proteins often form, and in a few places the tubes packed against each other in the way that these helices do in proteins. I wrote to Brian and told him what I'd seen and went to bed.

I was not prepared for the surprise that awaited me when I went into work the next morning. Of course, I had expected the usual emails from Utah telling me what had gone on while I was sleeping, but that morning there were several emails from Brian. The first was telling me that, yes, he could see that we had missed the density for proteins, and he'd identified one of them, S6.

At the resolution we had obtained, we couldn't build a new protein from scratch, but if a structure was already known, we could place it in the map so it roughly fitted into the density. Anders Liljas had collaborated with Maria Garber to solve the structure of S6 by itself. Proteins typically consist of structural elements like alpha helices, which should look like tubes at this resolution, and extended beta strands that go back and forth to form a flat sheet. Brian could take the atomic structure of S6 and place it in the density in the 30S by simply aligning where the tube-like helices and the flat beta sheet went. For the first time, we could see directly where a protein was in the 30S and how it was interacting with the RNA. It was like knowing in detail what an isolated steering wheel looks like and then for the first time realizing where it belongs in a fuzzy image of an entire car.

But that wasn't all. Over the course of a single night, Brian had located one protein after another, until he had located all seven previously known protein structures in the 30S maps. Actually, although he knew where it was, he had left S5 for me to place since he knew it was the first protein structure I had ever solved and had a particular affection for it. Placing a protein in the 30S subunit had been so exciting that Brian said it had been like eating crisps. Once he did the first one, he couldn't stop.

A lot of parts were now being placed into the fuzzy picture of the assembled machine. Initially, when we saw an RNA helix, we didn't know which part of the ribosomal RNA it was. With several proteins in place, it became possible to identify the pieces of RNA next to them because there were lots of biochemical data from people like Harry and Richard Brimacombe to show which proteins were near which pieces of RNA. Luckily, Brian had a lot of the data in his head or remembered where he could find it. So pretty soon, he had identified the segment of RNA near S6 and worked outward from there until he could see how an entire section of RNA called the central domain was folded. This was a much bigger breakthrough than we had ever dreamed of at this stage. We were seeing the molecular architecture of about a third of the 30S subunit, with the proteins and RNA all connected with each other to form an intricate complex structure. 'Looks like an article not a letter,' was Brian's verdict. Indeed, we had to start writing up our results right away.

The breakthrough came not a moment too soon. The next ribosome meeting was being held just north of Copenhagen in June, only a month later. When I contacted them before their deadline while I was still in Utah, we didn't have anything concrete, but we knew we had good crystals and were about to collect some data. In the worst case, I could simply say we'd entered the fray and had made some progress. On that basis, I asked one of the organizers, Roger Garrett, if I could give a short talk. The abstract or summary I submitted was a vague one-liner saying we'd report our progress on the structure of the 30S subunit. It said absolutely nothing beyond the equally vague title. Despite my vagueness, Garrett generously tacked me onto the end of the opening night's session on the structures of the entire ribosome or its subunits. He later told me that when the conference committee met, Peter Moore was doubtful that we had anything substantive since he had just seen me at Brookhaven a couple of months earlier (just before we began collecting our data). That was true: I didn't have anything then. But Garrett said he wanted to take a chance and include me because he somehow got the feeling I had something up my sleeve. We never imagined our progress would have been so rapid.

The next few weeks were a frantic effort to make sense of our results and write a coherent story for *Nature*. We wanted to send our paper off before I went public, because once the word was out, there would be a mad scramble with everyone trying to get their papers out. This is where being apart was not so helpful. In cases like this, different people are doing complementary tasks like making figures or writing different sections. They often have different schedules and don't always see the others doing their work. With the stress of a deadline, each often feels the others aren't pulling their weight and tempers fray. Normally I'd nip these tensions in the bud, but I wasn't there and had to calm everyone down over long distance. Somehow, we got the manuscript written, and I mailed three hard copies off to *Nature* just before setting off for Denmark.

The train from Copenhagen Airport to the quaint small city of Helsingør took about an hour. Just across the strait from Sweden, its English name is Elsinore, of *Hamlet* fame, and the play is performed there every year. Steve White and I were sharing a room. I had told Steve a long time earlier about our efforts to start working on the 30S. Like many others, he had been somewhat sceptical about the enterprise. When we had located the seven proteins in the 30S maps, I realized I had to let him know right away because he would be talking about their structures in isolation. This meant that his talk on our work on individual proteins would come across as rather anticlimactic. So I had called him and let him know. But that was before we had solved the architecture of an entire domain of the 30S subunit, showing how the RNA was folded and how the proteins bound to it. When we met in Denmark, I told him the full extent of our work, and he seemed a bit shocked by our progress. But ever the gentleman, he was gracious about it. Not everyone would be.

Most of the participants dribbled in during the afternoon, and there was to be an opening session just before dinner. The speakers were Tom, Ada, Jamie, and me, in that order. Except for a very few people, this was to be the first time anyone had any inkling that I was working on the entire 30S subunit. Suddenly, I was more nervous than I had been in years.

The talk began with Tom describing how they had made progress on the 50S structure. They were now at a similar resolution as we were and could see similar features in their RNA and proteins. They had also identified a few proteins and elements of RNA. After his talk, Ada got up and tried to make the point that he shouldn't be seeing proteins at all because the crystals of the 50S subunit were grown in high salt. Her point was that the electron density of a concentrated salt solution would be so high that the proteins would have very little contrast and shouldn't be visible. Tom disagreed with her, and she persisted. Finally, he lost his patience, folded his arms, and said, 'I've solved lots of structures in 3 molar ammonium sulphate. How many structures have you solved?' There was an awkward silence, and that was the end of the discussion.

Ada then got up to speak. She reported progress on her new 30S crystals and showed maps, including a couple of proteins. But unlike Tom's maps, we weren't shown the sort of clear features and overall shape that suggested that there had been a breakthrough.

Next, Jamie Cate got up to speak about their progress on the entire ribosome. Given his history, I had expected he would use osmium hexammine to phase the data using MAD. Actually, he had used the very similar iridium hexammine, which he told me later was a lot easier to make. The maps were at about 7.8 Å resolution. This meant he couldn't see the architecture of proteins that the Yale or our groups could, but he very clearly showed the grooves of the RNA double helices, including the long helix we saw running down the face of the 30S. The most interesting thing in his talk was that we could see the shapes of three tRNAs nestled in between the two subunits and the entire ribosome in greater detail than ever seen before.

Then it was my turn. I started off by saying how we had managed to get crystals of the 30S subunit that diffracted well without having to stabilize it in any compounds, simply by carefully purifying it and removing a protein component that was only partially bound so the 30S subunits were homogeneous. Then I described our progress in getting maps we could make sense of. I said that letting Brian have a crack at these maps was like giving a teenager the keys to a Ferrari – a remark that was later quoted in *Science*, leading his mother to call him Ferrari Boy. I had seen Harry sitting in the front row, and this was also a dig at his fondness for Ferraris. Finally, I talked about how Brian had figured out the architecture of an entire domain. There was complete silence after my talk, before Anders Liljas, who chaired the session and must have been taken aback by our unexpected foray, asked me how long we had been working on it. Ada asked me how we phased the structure. I didn't want to give too much away because the structures were far from being done, but I gave her a general outline of the heavy-atom clusters and other compounds we had used. This reticence was foolishly paranoid since Jamie had already gone public with their strategy, which was identical to ours. But it is hard to change one's mind-set, and I still had the mentality of someone trying to catch up from behind. There were a few more questions, and then it was over. We had not only caught up with the field but had also, at least temporarily, built more of the ribosome than any other group.

There was a huge buzz in the room as people started talking animatedly on their way to the dining hall. It was immediately clear to everyone that, after forty years, the ribosome field was about to change dramatically. Lots of people complimented us on our work, and Tom was congratulatory but seemed a bit rueful that I hadn't let on anything at all, even just before the talk. Peter seemed proud of his protégé. Harry looked very thoughtful and also very surprised at our apparently having come out of nowhere.

Not everyone was happy. Many biochemists, who were trying to get at structure using biochemical tools, suddenly realized their way of life was about to end. The foremost of these was Richard Brimacombe, who like Harry had been one of the few ribosome biochemists to dedicate much of his life to working on the RNA part of the ribosome. An Englishman who had spent most of his career at Wittmann's institute in Berlin, he had become interested in the problem of translation when working on the genetic code with Marshall Nirenberg in the 1960s. More recently, his idea was to combine the painstakingly obtained biochemical data that he, Harry, and others had generated on roughly which proteins were near which piece of ribosomal RNA with low-resolution blobby images that were being generating by electron microscopy. The electron microscopy maps he was working with were generated by Joachim Frank's rival, Marin van Heel. The problem was that the data were not precise and the images not sufficiently detailed to recognize features unambiguously. Still, for a while, it seemed as though this would be the only way to get some sort of approximate molecular model of the ribosome. Oddly, he got closer to the truth with the more complicated and much bigger 50S subunit than with the 30S. But after that night's session, he could see that he was soon going to be steamrolled by progress towards a high-resolution structure of the ribosome. It could not have been easy for him to give his talk later in the meeting.

Peter Moore could sense the anxiety on the part of many in his meeting. In his summing-up remarks that marked its close, he tried to assuage their fears by paraphrasing Churchill: 'This is not the end. It is not even the beginning of the end. But it may be the end of the beginning.' Certainly, for those biochemists who were focusing on how the ribosome worked rather than what it looked like, that turned out to be true.

It could not have been easy for Ada, who had started the crystallographic work about two decades earlier, to suddenly see others making such rapid progress and overtaking her. I heard from several people at the meeting that she was less than charitable about Tom and me. I could see how she might feel about the turf she had so long occupied by herself suddenly being invaded from all sides, so when it came time to contribute a chapter to the book on the conference, I thought it would be appropriate to start it off with a description of her breakthrough on the 50S crystals and how that pioneering work had paved the way for what came later. It turned out to be the opening paragraph in the opening chapter in the book, but if I'd naïvely expected that this homage would mollify Ada and lead to peace and goodwill all around, I was soon disappointed. The next couple of years turned out to be a period of fierce and acrimonious competition.

No matter how amazing they seemed, the results so far, even from the Yale group and us, were only progress reports, and what we could do with the maps at this stage was limited. If we already knew the structure of an isolated protein, we could place it into the maps, but the resolution was not good enough to build those parts of the ribosome for which there was no prior information. So we could not really figure out the structure of something completely unknown. Moreover, even the models for the parts we could construct were only approximate. They could tell us the rough architecture but not the detailed and accurate atomic structure that would allow us to understand its chemical mechanism. Now that we had shown that the crystallography was working, it would be a flat-out race to get data to better than 3.5 Å resolution and then build a complete atomic model of each subunit.

To Peter, this seemed inevitable. At the conference, I met Ditlev Brodersen, who had agreed to join my lab. I introduced him to Peter and said he was going to join my lab as a postdoc. Peter wryly asked him if he was good with RIBBONS, which at the time was the leading computer program to display structures of proteins and RNA. He was joking that the structure would be done so quickly that all that would be left for Ditlev to do would be to make the figures for the paper. Unfortunately for us, it turned out not to be the case – although Ditlev turned out to be a master of lots of skills, including RIBBONS.

CHAPTER 12

Almost Missing the Boat

BRIAN AND I RETURNED FROM Denmark ecstatic. Soon afterwards, a conference on nucleic acids was held in America. Neither the Yale group nor I were there, but Harry was. Brian, who went to the meeting to speak about our work, gave me an entertaining account of the proceedings.

Harry was the subject of a great deal of adulation. There was already a lot of speculation about an eventual Nobel Prize for the ribosome. One young woman at the meeting asked Harry if he would take her to Stockholm with him and had him autograph her badge. Brian also told me that during a meal, Harry said, 'The trouble with Venki is that he hasn't been doing this long enough.' Long enough for what? To be considered one of the leaders? I thought it was a curious comment considering I'd been working on ribosomes ever since I was a postdoc, or over twenty years at that point. It was clear that even before the structure was actually solved, politics had begun to infect the race.

Our submission to *Nature* was sent to the usual three anonymous reviewers. One of them was Steve Harrison, who was excited enough about the manuscript that he not only gave us detailed suggestions on how to improve it but disclosed his name. Some of his comments were a little paternalistic, like crossing out a paragraph where we described the reasons for the breakthrough, saying, 'Don't brag!' Perhaps we had got a little carried away with our own success.

Word had got around at the Denmark meeting that we had already submitted a paper to *Nature* on our results. Exactly as I had anticipated, this set off a mad scramble from the other labs to get their stories out around the same time. Soon afterwards, the Yale group submitted their story on the 50S to *Nature* also; the journal held ours so that the two could appear together. Although I was initially annoyed not to have our special, unshared moment in the sun, in hindsight the two papers made a nice pair, one on each subunit. The papers made something of a splash, with news items about them in several journals.

Harry, too, got busy, and soon afterwards submitted his paper on the entire 70S ribosome to *Science*. It isn't just scientists who are competitive, but journals are too. *Science* published his paper rather quickly with a great deal of fanfare, including an accompanying piece on the ribosome race by a journalist, Elizabeth Pennisi. She had written a short piece on our *Nature* papers only a month earlier, in which she had mentioned Ada and Harry mainly in passing. This time, she wanted to do a longer piece and started talking to various people. I suggested that if she wanted to talk to someone outside the ribosome field who was respected, she could talk to Steve Harrison.

Steve, however, thought Pennisi's treatment of Ada in her earlier piece was dismissive and refused to talk to her further, saying she should talk to Ken Holmes, a famous biophysicist at a Max Planck institute in Heidelberg. Ken had changed the world of biological structures forever by showing that the intense X-rays produced by synchrotrons, which were used mainly for high-energy particle physics, could be used for diffraction studies. He and Gerd Rosenbaum, who, when I met him a year later, struck me as a brilliant if somewhat opinionated and eccentric specialist in X-ray optics and instrumentation, built the first X-ray diffraction beamline at the

DESY synchrotron in Hamburg. This was the same synchrotron outside which the Max Planck Society had conveniently built a lab for Ada and where she had collected much of her early data. Having watched Ada work on the problem for years, he was not entirely charitable to us. 'She's done most of the backbreaking work,' he was quoted as saying. 'The others have jumped the gun. They should have left Ada in peace.' I was shocked and dismayed – it confirmed my worst fears of how some would perceive our efforts. Luckily for us, most people were happy about the breakthroughs and very supportive. And just two years later, when I first had the chance to meet Ken in person, he was very friendly and complimentary about our work, and we have remained on good terms ever since.

Pennisi's piece also quoted Joachim Frank, referring to me as the dark horse of the Denmark meeting. He had meant to be complimentary, implying I had come out of nowhere and surprised everyone. But coming in the same report as Ken's remarks, and not long after I'd heard about Harry's comments, I found myself irrationally annoyed by it, seeing it as yet another effort to belittle me as a Johnny-come-lately. Of course, since I had deliberately kept our efforts very quiet, because we were starting out from behind, I could hardly complain about the dark horse description. But it caused a lot of ribbing by my colleagues – including some joking references to my dark skin. One friend superimposed a picture of my face onto the head of a black horse and said I could use it as a logo in the slides of my talks. Another idea was to shrink it down to the size of a dot on the corner of the slide and have it gradually grow in size as the slides progressed so that it would become obvious as a horse's head only towards the end of the talk – a homage to the famous scene of Omar Sharif's entrance in David Lean's *Lawrence of Arabia*.

Suddenly being thrust into the scientific limelight was a new and uncomfortable position for me. It was also a distraction I could not afford. The real deal – the complete atomic structure – was still ahead of us. Getting the first atomic structure of a ribosomal subunit would now be a flat-out race between Yale and us, or so I thought. After all, the 70S crystals were not of sufficient quality to yield an atomic structure, and Ada seemed to have been left behind.

In any case, it was time to get back to work. We needed new crystals and high-resolution data. The rest of the lab flew out to Brookhaven from Utah with more crystals, and I joined them from England. Now that we knew which compounds were going to be the most useful, we thought we could focus on just those to collect more data that might give us higher resolution. But even with this focused strategy, our trip led to almost no improvement over our previous maps. It was memorable mainly for my accidentally locking us out of both rental cars at a shopping mall in Shirley just south of the lab at Brookhaven and having to wait a couple of hours for the company to rescue us. Clearly, we needed to start again in England.

While I was away at Brookhaven, Vera had found and successfully bid on a house in Grantchester, a picturesque and historic village just three miles west of the lab, so at least I could stop worrying about housing and concentrate on work. At the LMB, I was given four lab benches in a room that used to be Aaron Klug's old lab. It was the same room in which I had spent my sabbatical with Wes Sundquist, and many of its previous occupants had gone on to fame and glory. That history weighed heavily on me as I began to set up my lab for the autumn, when everyone would arrive.

I advertised for a technician and gave the job to Rob Morgan-Warren, an earnest and muscular young man who had a black belt in martial arts and had just graduated from Birmingham. He and Andrew Carter were the first to arrive, and I took them through the basics of making ribosomes. Andrew and I had to figure out how to use a new chromatographic system, and he soon took charge of it. Around this time, Bil arrived from Utah and met Andrew and Rob.

The contrast between Bil and Andrew could not have been greater. Andrew was from an academic family and had gone to Winchester, the oldest public school in the country—public in the British sense, meaning not public at all but very private. From there he had gone to Oxford to study biochemistry and then come to the LMB to start his PhD. Not only was his elite education quite different from Bil's but so were their interests and personalities. Although Bil was an experienced senior graduate student with a track record behind him, Andrew was intellectually very confident and not the sort who would kowtow to Bil – or anyone else for that matter. Bright people tend to have strong personalities. With Bil, Brian, and Andrew, I could see that making sure the people in my team got along and worked together would be almost as hard as solving the 30S subunit.

Brian's summer was exciting scientifically but tumultuous personally, having gone through a divorce right in the middle of it. He briefly wondered at one point whether he should even join me in Cambridge. In fact, both Bil and Brian had done enough in Utah to finish up and move on to the next phase of their careers, so it was a very good thing for me that they decided to move to Cambridge to help finish the 30S structure. I was relieved when Brian finally showed up, looking a bit tired and worn by the events of the summer but ready for the next phase.

The last to arrive was Ditlev Brodersen. I had met him briefly in Denmark, but the meeting had been so hectic that we had not had a lot of time to chat. He turned out to be not just smart but versatile, superb at everything from computing to lab work. He was also friendly, pleasant, and good humoured, traits that would be sorely tested in the next year.

With the team in Cambridge assembled, it was time to get cracking again and produce more crystals. It is the nightmare of every crystallographer that moving to a different lab and using different sources for chemicals or even a small change in the temperature of the cold room could suddenly mean that nothing works. Fortunately, we were able to reproduce decent crystals, which Bil and I took to Brookhaven, but again, there was little improvement.

A serious difficulty was that compared to the 50S crystals, the diffraction spots from our 30S crystals at the high angles that we needed to get to high resolution were very weak. To make matters worse, the beamline at Brookhaven spread out these high-angle spots over a larger area, making their intensities even harder to measure. To measure the weak data accurately, we had to expose our

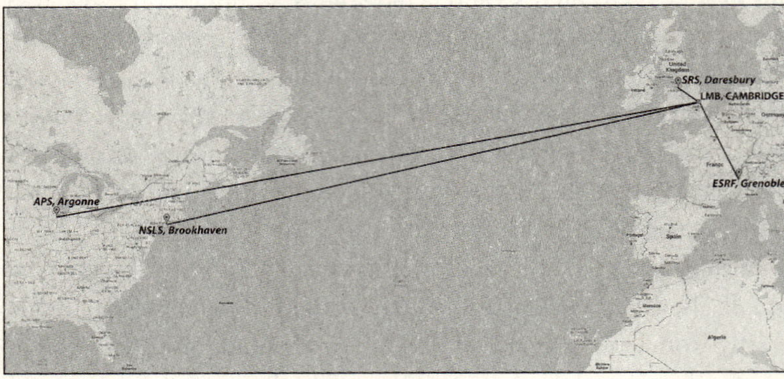

Figure 12.1 Join a ribosome lab and see synchrotrons around the world
(map data © 2018, Google, INEGI ORION-ME)

crystals much longer. This meant that even at the low temperatures used for crystallography, the crystals still suffered damage while we collected the data and we could not improve the resolution much. We had reached the limit of what we could do at Brookhaven.

Closer to home, the LMB had routine access to two synchrotrons, an old one in Daresbury in the north of England and a newer, much more intense one, the European Synchrotron Radiation Facility, or ESRF, in Grenoble, France. We quickly figured out that we could not collect any better data at Daresbury, but the synchrotron turned out to be very useful for checking the quality of our crystals. The one in ESRF was potentially great. It was a relatively new beamline and was going through teething problems. But that was not the most serious issue. The high intensity on the Grenoble beamline meant the crystals died even faster. This meant we would need lots of crystals to get a complete set of data from each of our heavy-atom derivatives, and if we wanted to do a MAD experiment, we would need not one but three complete data sets from each soaked compound. Moreover, the crystals were quite variable. Not all of them diffracted well enough to collect high-resolution data, and we didn't quite understand why. Even for the ones that diffracted

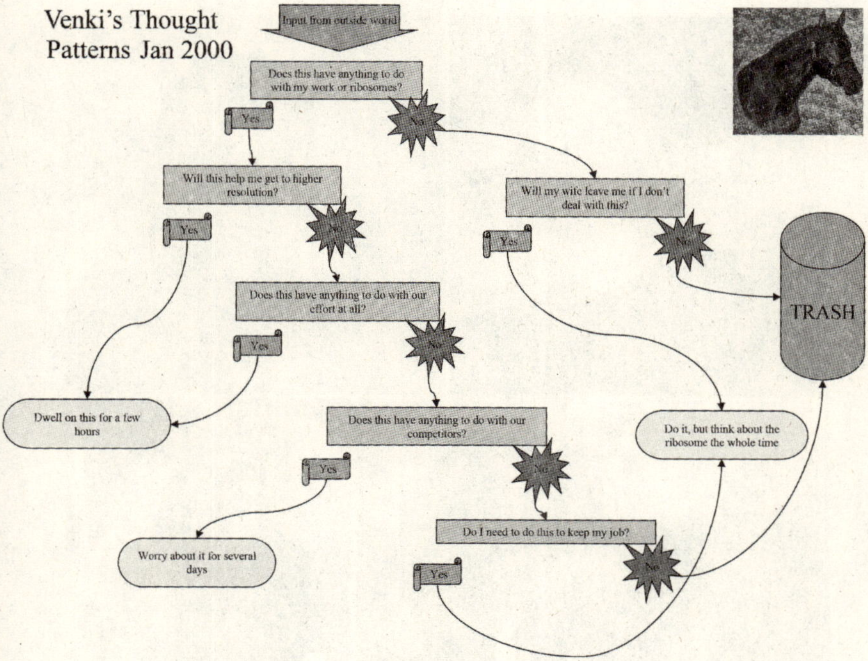

Figure 12.2 Flowchart of the author's thought patterns in 1999–2000
(courtesy of Bil Clemons)

well, the repeat distance or unit cell was slightly different from one crystal to the next.

We were stuck. We couldn't collect high-resolution data from just one crystal because it would be damaged long before we could collect a complete data set from it. Because the crystals were so variable, we couldn't combine partial data from lots of crystals to generate a complete high-resolution data set. This meant that we would not be able to improve on the resolution we had. My euphoria from Denmark vanished. For the next few months, I was obsessively focused on finding a way out of our predicament.

So much so, that one day I found that Bil had made a flowchart and pinned it on the door of the lab. It was titled 'Venki's thought patterns, Jan. 2000,' with a picture of a dark horse in the right corner.

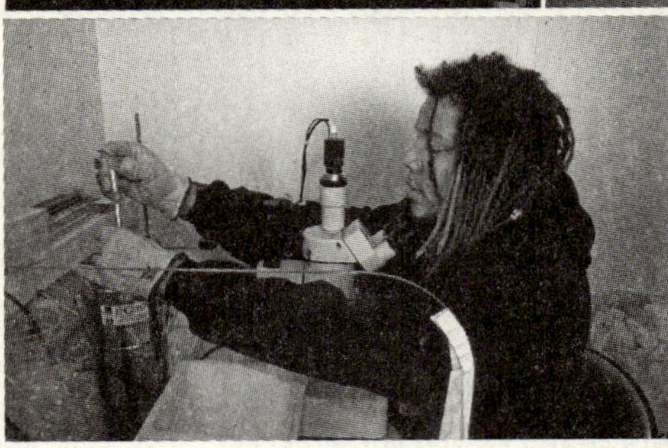

Figure 12.3 *a.* Ditlev Brodersen and Andrew Carter, *b.* Bil Clemons freezing crystals in the cold room, and *c.* Bil Clemons and Rob Morgan-Warren with hundreds of empty vials and canes at the end of the experiment

To their credit, Bil and Ditlev were willing to try all sorts of new ideas to get around the problem. Radiation damage is still not entirely well understood, but essentially, two types of damage occur: primary and secondary. We can do nothing about primary damage, in which a bond is broken due to an X-ray knocking an electron out of its orbit. But I thought we could try to minimize secondary damage, in which some of the damage to molecules generates highly reactive free radicals that diffuse and cause further harm. My knowledge of hard-core chemistry was pretty minimal, and I frantically thought of compounds to add to our crystals that might scavenge free radicals before they damaged our ribosomes. Ditlev tried a few like ascorbic acid (vitamin C), but none of them worked.

If we couldn't slow down radiation damage, perhaps we could make the crystals more uniform so that we could use lots of crystals, each yielding a small fraction of the data before it was too damaged. This led to a crazy idea. What if the crystals had all started out the same, but variable rates of freezing meant they expanded or contracted to different degrees and ended up different from each other?

By now, Bil had developed a routine for the incredibly tedious and unpleasant job of freezing lots of crystals in the cold room. He would put on his jacket and go into the cold room, get all his equipment set up, then set up a mini-stereo system, put on his favourite Johnny Cash CD, and for the next several hours, with hardly a break, manually fish out one crystal after another with a loop attached to the end of a pin protruding from a magnetic metal base, and then plunge them quickly into liquid nitrogen and store them in vials. Rob would be his assistant throughout the process. After a session like this, they would have dozens of vials filled with liquid nitrogen, each with a crystal in its loop. The vials would be slotted into metal canes that could hold four or five crystals, which would all be stored in a dewar (a vacuum flask) containing liquid nitrogen, waiting to be shipped off to a synchrotron. Even though this was done mostly by one person, Bil, maybe the process was too variable.

When my colleague Phil Evans heard about my problem, he showed me a device that worked a little bit like a guillotine. After fishing out the crystal from the drop, you attached the assembly (with the loop containing the crystal pointing downward) to the top of the guillotine, below which was a little container of liquid nitrogen. As soon as you pressed a pedal, the guillotine would drop, so the crystals would all be plunged into the liquid nitrogen at exactly the same speed and perfectly vertically every time.

Bil agreed that this might be the answer to our problems. If this idea worked, it would get us over the hump. With his usual enthusiasm about trying anything new, Bil took a couple of hundred of the best crystals, that had taken about eight weeks to grow, soaked them in the various heavy-atom compounds, and used the guillotine to freeze them all.

Over the weekend, Bil and the team took the crystals to the Daresbury synchrotron. That Sunday morning, I got a phone call from Bil. 'You know, boss,' he began (he always ironically called me boss), 'the French Revolution was not a good thing.' I had absolutely no idea what he was talking about. It turned out that the guillotine had never been used for freezing crystals. It was actually a device for plunging grids for electron microscopy. So each time Bil had used it to freeze a crystal, the guillotine had plunged down and stopped with a thud, and the crystal had shot out of its loop to disappear forever into the container of liquid nitrogen. Virtually every single loop they had looked at was empty, except for one in which the crystal had been frozen just before it could detach from the loop. With the loop looking like a hand, the long crystal protruding from it looked like it was giving Bil the finger.

We had just lost two hundred of our best crystals. More importantly, since the crystals took so long to grow, we had set ourselves back by at least two months in the middle of a tight race – all because neither Bil nor I had thought it would be worth checking the guillotine device out with a couple of crystals first before trying it with all of them. I spent the rest of the weekend in a state of shock. But there was nothing to do but start again with the arduous job of making more 30S subunits and crystallizing them.

While we were waiting for more crystals to materialize, I also realized that even if we got lots of matching crystals, we would still have a problem. Our whole strategy for obtaining phases depended on anomalous scattering from special atoms like osmium, which gave rise to a small difference in intensity between symmetry-related spots called Friedel pairs. To calculate a structure, we had to measure those small differences very accurately. If we collected data on crystals that were not precisely lined up symmetrically with respect to the beam, the two spots in the pair would be measured at different times, so one spot would have suffered more or less radiation damage and we would have to apply a correction.

Or, since our crystals only lasted a short time in the beam, the two spots might often be measured on different crystals, and since the crystals varied in size and shape, we would have to correct the data from each to put them on a common scale. Either way, the errors in the corrections could themselves be much larger than the small differences we were trying to measure.

A way to get around this problem is to line up a crystal very precisely along one of its axes of symmetry so that the diffraction pattern would look symmetric, with the pattern on the left side of the detector looking like a mirror image of the right side. This way, when it is rotated, the symmetry-related spots on either side would show up at exactly the same time. By measuring both spots in a pair from the same crystal at the same time, we would eliminate a lot of the errors that come from compensating for different crystal sizes and radiation damage.

When you fish a crystal out with a loop, you are lucky if it sits in your loop at all. There is no way to control its direction or even position in the loop. On most instruments, you can centre the crystal in the beam, but there is no way to line up the crystal so that the diffraction spots would be symmetric. However, if you have an X-ray instrument that can rotate the entire assembly containing the crystal in its loop and metal base about different axes, you can first take a few test shots from which you can calculate the precise orientation of the crystal relative to the apparatus, and then use the motors that control the various axes of the instrument to line up the crystal perfectly about the beam regardless of how it originally sat in the loop. We had used this device, called a kappa goniometer, at Brookhaven to do exactly this. The problem was that although the ESRF beamline in Grenoble was steadily improving, it didn't have one.

I thought the answer to our problems was the beamline at the Advanced Photon Source (APS) at Argonne National Laboratory just outside Chicago, which had both the high intensity of the Grenoble beamline and a kappa goniometer. It was meticulously designed by the same Gerd Rosenbaum who, working with Ken Holmes, had built the first X-ray diffraction beamline at the DESY synchrotron in Hamburg. In the middle of October 1999, as soon as I realized we needed it, I wrote to Andrzej Joachimiak, who oversaw the beamline, to ask if we could get some time on his machine. It had just become open to the public, but even before that, in the 'testing phase,' I knew Ada had collected lots of data there. I didn't get an answer, so I wrote again in early November. He wrote me a brief reply saying that he would get back to me over the weekend and the earliest time I could get was in the beginning of the next calendar year. I didn't hear back from him, so I tried again in late November.

November turned into December, and I kept getting reports that the Yale group was slowly pulling ahead. Finally, in mid-December, I heard they'd had a little celebration on reaching 3.1 Å resolution. This meant they had cracked the problem, and they would be able to build an atomic structure into the maps. Finally, I'd heard they got even better data at the APS.

New Year's Day of 2000 was a big worldwide celebration of the millennium, but I was despondent about our lack of progress. I felt like someone who had run a few good initial laps in a race but was now falling inexorably behind. On January 3, Peter wrote to say he was giving a talk in Manchester and asked if I would like him to visit us in Cambridge to talk about their structure. Much as it hurt, I could hardly say no – it would be of tremendous interest to everyone, not just my group. I used the opportunity to ask him about the APS, and he said the data they had collected there were much better than anything they had obtained elsewhere. Right away, I knew that I had been right to ask for time there. I told Peter that I was stuck, and we had asked for time on the APS instrument as far back as mid-October but had still not obtained a response to our request. Paul Sigler – the man who was pushed off the stage prematurely at the Seattle crystallography meeting four years earlier – was Peter's colleague at Yale. I knew he was a member of the committee that oversaw the APS beamline and also that Andrzej had once been his postdoc. So I asked Peter if he could put in a word for us with Paul. I also wrote directly to Andrzej again and left a couple of messages on his phone.

The next day, January 5, I got a reply from Peter saying he would talk to Paul, which, considering its effect on me, sounds as biblical as it should. The following day, January 6, Peter wrote saying that Paul had contacted the APS and hopefully they would be in touch shortly. The very next day, there was an email from Andrzej apologizing for the delay in replying to me, saying he had been very busy, but they could give me time in late March. That was almost three months hence and would be six months since I had originally contacted Andrzej. It meant we had lost six months at a time when our chief competitors were using his beamline to get ahead. Still, it was an offer I couldn't refuse.

Four days later, on January 11, I received shocking news. Paul had died of a massive heart attack on his way to work. Paul's death was a tragedy in many ways. He was a great figure in structural biology who had trained many superb scientists, and at the age of sixty-five was still relatively young, with many years of science ahead of him. The whole episode saddened me. But it also made me realize, yet again, how science depends on weird quirks of fate. If I had not written to Peter just the previous week, and Paul had not immediately interceded on our behalf, I am not sure Andrzej would have responded in a timely manner, and we might well have been completely out of the running for the first atomic structures.

This was brought home to me soon afterwards when, a couple of weeks later, I had to give a talk at the ESRF synchrotron in Grenoble at a meeting for their users, and both Ada and I were invited to talk. I wouldn't have shown our latest research anyway, but it didn't matter as I had nothing new to report. So I gave a talk based on what we had already published a few months earlier. After my first slide, I heard a click and turned around to see someone from the audience taking a photograph. I thought he was possibly reporting for the synchrotron newsletter, but then I heard a click after the next slide. And then after every single one. The guy turned out to be someone who had worked with some of Ada's group in Hamburg. When I wrote to him and said I'd be happy to send him a copy of my talk, he said he had been 'asked by former colleagues to give a small report' about my talk. It is considered very bad form to take pictures of someone's talk without their permission. They must have felt the stakes were high enough to get someone to do this, but in this case, they didn't get a thing from me that wasn't published.

In an amusing coda to this, a year later, when the structure was finished and published, I gave a talk at the same venue. As soon as I started, a young woman took a picture. Thoroughly irritated, I stopped and said there was no need to take any pictures, and she could have a CD of my talk if she wanted. The ESRF director came up from the front row and apologized to me, saying she was just the photographer for their newsletter!

In any case, right after my talk, Ada spoke, and it immediately became clear that I had badly underestimated her tenacity when I had ruled her out of the competition. Her maps had improved considerably from the previous summer in Denmark and were now actually better than anything we had. Clearly, whatever problems she may have had earlier, she seemed to be making progress now. Her maps still weren't good enough to start building an atomic structure, but there was no telling how long it would be before she got to that stage. Much of her new data had been collected at the APS on the same instrument where we had finally obtained some time.

I went back realizing that our two days at the APS would be absolutely crucial. We could not afford to fail, and we could not leave anything to chance, since otherwise we might never be able to catch up. In a heroic effort, Bil, with help from Rob and Johnny Cash, froze hundreds of crystals and stored them away in dewars. We had created a spreadsheet so each crystal could be located precisely in its dewar. Every other one was screened at the Daresbury synchrotron for quality and cell dimensions. The ones that had matching cell dimensions and diffracted well were considered keepers and gave us an idea of how many usable crystals we had for each heavy-atom soak. Not content with this, I persuaded Bil to go to the Brookhaven synchrotron by himself and collect actual data from representative crystals for each compound. After an exhausting forty-eight hours without sleep, he obtained data showing that all the compounds were bound as expected. With three dewars full of decent crystals, we were ready for our make-or-break trip.

CHAPTER 13

The Final Assault

THE END OF MARCH 2000 had finally arrived. Bil, Ditlev, Rob, and I arrived at Heathrow with the three dewars full of crystals. One of them was in a rectangular case, and the other two were in cases that were cylindrical with dome-shaped caps, resembling what Brian called a small thermonuclear device. Among ourselves, we referred to them as the suitcase and the bombs and had to consciously refrain from that habit at the airport when checking them in as baggage (today no airline would accept them as baggage and they now must be shipped separately by FedEx). We certainly didn't want a zealous airline or security official opening the dewars and warming up our precious crystals.

It was an icy cold day when we landed in Chicago, rented a car at O'Hare International Airport, and drove down to Argonne where the APS was. After a fitful night's sleep, we assembled at the Structural Biology Center (SBC) beamline, where we were met by Andrzej and his colleague Steve Ginell. Beamlines are incredibly complicated, and each one is different from the others, so outside users need all the help they can get from the local staff. Our session was to begin with instruction on various safety rules and an overview of the beamline.

Figure 13.1 Picture of SBC crystallography beamline at the APS synchrotron in Argonne. The X-ray beam comes in through a pipe from the right (1) and hits the crystal (2). The diffracted X-rays are measured using a CCD detector (3). A cooling device (4) keeps the crystal bathed in a stream of very cold nitrogen gas, which is obtained from a large liquid nitrogen storage dewar flask (5) *(courtesy of Andrzej Joachimiak, Argonne National Laboratory)*

I was a little apprehensive about how welcome we would be. Perhaps Andrzej had been unresponsive for months because he felt a loyalty to Ada, who had been collecting her data there, and he didn't want to help a competitor. Moreover, he might have felt coerced by Paul into responding.

It soon became clear that my fears were completely unfounded. After the typically thorough and seemingly interminable induction procedure – which was coming out of our precious forty-eight hours – we started collecting data. The first diffraction image was better than anything we had seen from our crystals before. Andrzej hung around to make sure our strategy would work. Almost nobody else had used the ability of their beamline to pre-orient crystals to get more accurate anomalous data, and it required them to

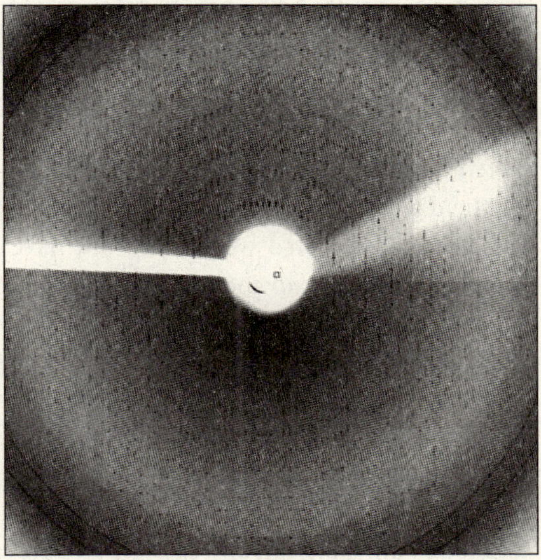

Figure 13.2 A picture of an aligned diffraction image from a 30S crystal at the APS at Argonne

run a separate program to do it. The first crystal went up; we took our test shots and told the program to orient it. Then we took another shot to see if it had worked. It was beautifully aligned: the spot pattern was now perfectly symmetric and extended all the way to high resolution. We were almost in business.

Almost – because I had made a rookie error in crystallography. To measure the data, you have to rotate the crystal a very small angle, say from 0.0 to 0.1 degrees, measure all the spots that satisfied the Bragg condition in that range, then go on from 0.1 to 0.2 degrees, and so on. I had estimated that 0.1 degrees would be fine enough a sampling to prevent two different spots from overlapping on the same part of the detector within a single image, but I hadn't accounted for the fact that real crystals are not perfect but contain tiny mosaic blocks that are all in slightly different orientations relative to one another. Each block would satisfy the Bragg condition at a slightly different point in the rotation. This meant that as we rotated the crystal, each spot would persist for a much larger angle than if the crystal had been perfect. Because of this, a spot that should have appeared later in the rotation after an earlier spot had disappeared would instead overlap with the first. With these overlapping spots, there would be no way to separate them and measure the individual intensities.

Fortunately, Andrzej immediately saw the problem and suggested we reduce the rotation for each frame – in other words, sample the data more finely. If he hadn't given us this tip, the fact that our crystals diffracted well would not have helped us: we would have not been able to measure complete data to high resolution, and our maps would have been much worse. It might have made the difference between solving and not solving the structure.

I immediately felt very embarrassed for having been suspicious of Andrzej's motives. Probably in the middle of trying to get the beamline out of the commissioning test phase and ready for general users, he had simply forgotten our requests and was jogged into action by Paul Sigler. In any case, he and Steve Ginell seemed quite excited by our data and were very helpful. Once we had to wake Steve up at some ungodly hour because the beamline had stalled, and we didn't know how to restart it. Andrzej also got Wladek Minor, one of his friends from the 'Polish crystallography mafia' to modify the program to measure the intensities because we were collecting many more frames and at a finer angle of rotation than the standard version could cope with. This program, HKL2000, was written by Zbyszek Otwinowski and Wladek.

Zbyszek, who had written the guts of the algorithm, was something of a genius. He had studied physics in Poland and on emigrating to the US had found a job helping out in Paul Sigler's lab, which was then in Chicago. He noticed that some lab members were calculating a Patterson map and said, 'That looks like an autocorrelation function.' They were amazed that this low-level help knew exactly what it was. Paul was so impressed by Zbyszek that he encouraged him to apply to grad school and get a PhD. He told me that he got a call from the organization that administers the GRE, the nationwide entrance exam for grad school, saying they suspected someone of cheating, and the student concerned had referred them to him. Apparently, Zbyszek had achieved perfect scores, something they said was highly improbable. When Paul found out it was Zbyszek, he laughed and told them not to worry. Zbyszek was also the person who wrote the program that I used to solve my first MAD structure and got me thinking about doing the ribosome.

When I saw all the special help we were getting while we were collecting data, it reminded me what an inbred world science often is. Like so many others, Paul had worked at the LMB too. Although I had known him for years, I had asked my own mentor Peter to ask him to put in a word for me with Andrzej, who was his protégé. Andrzej and Zbyszek knew each other not only because of the Polish connection but because they were both protégés of Paul's. It also made me realize how difficult things are if you are outside this inner circle where the action is, and how difficult it can be to claw your way into it if you are an outsider.

Anyway, we were on our way. To make the most of our forty-eight hours, we planned an almost military strategy. Each of us had twelve-hour shifts, but staggered by six hours so at all times we would always have someone experienced in crystallography and someone who wasn't too tired. I decided to take the 3 a.m. to 3 p.m. shift so I would just stay on British time.

The data were pouring in so fast that we had trouble keeping up. We absolutely needed to process the data on the fly to know which crystal to put up next and which part of the data was still left to collect. This is when I really appreciated Ditlev's superb computing skills. He would set up complicated computer scripts so that just changing a few keywords would send the next batch of data to be crunched away by the computers.

After an exhausting forty-eight hours, we had come to the end of our session. It was time to check if the heavy-atom peaks were present – that would tell us whether the experiment had worked. When the calculations finally spat out the peaks, they were barely above noise. My heart sank. Somehow, we had screwed up the experiment, and our make-or-break trip had been a failure. There was a stark silence in the room. Out of sheer desperation, I checked the code again and noticed we had made a mistake in the input and redid the calculation. There they were: the highest peaks were twenty-five times above the background noise, and there were lots and lots of them. More than we had ever seen before. The tension of the previous year suddenly dissipated. I got up and started dancing around the room saying, 'We're going to be famous!'

Only a few days after we returned to Cambridge, we had beautiful detailed maps of the 30S subunit in which we could clearly distinguish the shapes of individual bases on the RNA and amino acid side chains of the proteins. It was time to solve the molecular structure. Luckily, we didn't have to start completely from scratch. Using the lower-resolution maps we had generated earlier, Brian had already figured out roughly how the RNA was folded in most of the subunit and used the biochemical data to identify where all the proteins were. Now we could work out the detailed atomic structure by building in every single amino acid of the twenty or so proteins and every nucleotide of the RNA in the subunit.

Building chemical groups into the required density involved spending day after day in a darkroom wearing stereo glasses in front of a special graphics terminal.

Those were the days before video games had made graphics hardware so cheap that you could even buy one for your home entertainment. So even a well-equipped place like the LMB only had about four terminals that were fast enough to move around a large molecule like the 30S subunit without being too jerky. When we signed up for all four terminals for days on end, our colleagues were pretty annoyed, but when I told them the urgency of our situation, they were more accommodating. In the end, we agreed to leave one terminal free for everyone else and dominated the graphics room for several weeks. We split up different parts of the 30S subunit among us, and each built his own portion of it.

While the molecule was gradually being built, we had collected an even better data set that was to tell us quite a lot about how the ribosome worked and how antibiotics would block it.

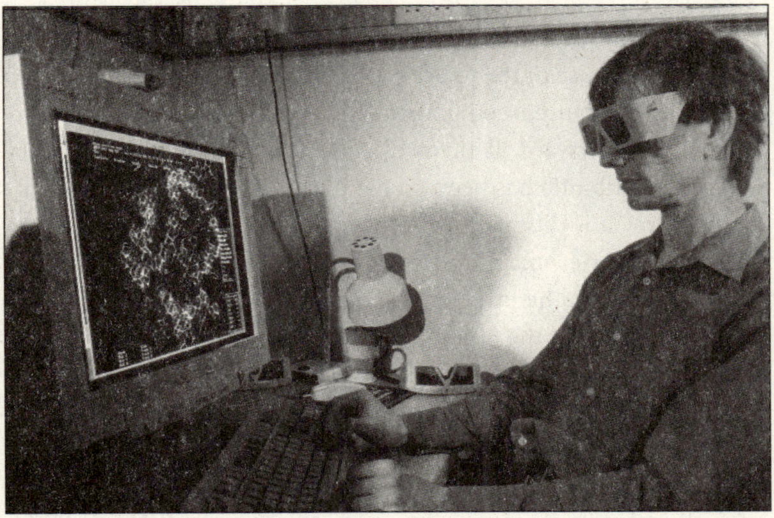

Figure 13.3 Brian Wimberly at a graphics computer. He spent most of his waking hours for a year in this dark computer graphics room.

This project started when we had problems with the variability of our crystals and we thought adding antibiotics would stabilize the crystals so they would not only be better in quality but more like one another. Andrew did a survey of lots of antibiotics that bound to the ribosome and read Harry's papers to see which parts of RNA were chemically protected by each antibiotic. He found a combination of three antibiotics that were thought to bind quite different parts of the ribosome and this yielded nice crystals that diffracted well.

Andrew and the team took these crystals to the ESRF synchrotron in Grenoble. By then, the instrument that had previously been a bit problematic was working well, so they ended up getting a very nice data set. Suddenly, we had data for the 30S subunit to which three antibiotics – spectinomycin, streptomycin, and paromomycin – were bound. These antibiotics had been known for almost fifty years, but nobody had seen exactly how they bound to the ribosome to get an idea of how they prevented it from functioning.

Solving the first structure of a large complicated molecule is quite hard because of the phase problem. But once you have solved that initial structure, seeing a small antibiotic bound to it is relatively straightforward. It involves collecting data from crystals in which the antibiotic has been added to the molecule, and then calculating how it is different from the empty molecule. The resulting 'difference Fourier' maps show you where the antibiotic is bound. Since in our case, the three antibiotics were bound simultaneously to different parts of the 30S subunit, we were able to see all three of them with a single data set.

I had fantasized about this moment for a long time. In my fantasies, though, we would savour the structure and think for months about what it all meant. In reality, we did not have that luxury because the other groups were close to putting something out.

Tom and Peter had already gone public with their structure, even if they didn't disclose all the details of it. As I saw in Grenoble, Ada too had made progress. Only a couple of months earlier, I had been scheduled to speak in Heidelberg at an international meeting where Tom was invited, but since we had not yet made our breakthrough to high resolution, I pulled out of it. I also declined to speak at a tRNA meeting where Tom spoke about his structure. This was particularly awkward because it was held right in Cambridge, but I thought our results would look very lame and 'last year' compared to the Yale group's progress.

Things were different now, and there were two important meetings coming up in July. One was the International Congress of Biochemistry, a triennial meeting, which that year was, by coincidence, in Birmingham. Another was a meeting Harry and his friends had organized in Santa Cruz. They were within a few days of each other, and all of the groups would be represented. If we didn't go to either, we would be perceived as also-rans, but I also didn't want to report our findings before we had submitted our papers. I rang up the editor at *Nature* and told him we had cracked the problem and when he could expect something from us.

Suddenly, in the middle of this frenetic activity, I got a strange letter postmarked Grenoble and signed 'Robert.' It warned me that he had seen Ada's and my talks in Grenoble earlier that year, and I should not speak in Birmingham because I would look bad compared to Ada; instead, I should wait until I had a complete structure. 'Robert' went on to say he saw I was not speaking in Santa Cruz, so 'there is no shame there.' The only problem was that I didn't know anyone in Grenoble named Robert and nobody in Grenoble had the slightest idea who might have mailed this letter. I showed the letter to my lab and other colleagues, and we all thought it was extremely strange; nobody had ever encountered anything remotely like it. But if it was meant to intimidate us in some way, it merely had the opposite effect.

While we were writing, I went to a meeting in Erice, a beautiful mountaintop medieval town in Sicily where they periodically hold crystallography meetings. I had been there once before and loved the place. Tom was supposed to be there too. By then, we had cracked the structure but hadn't written our papers yet, so I didn't want to let on any more than we had to. As it turned out, Tom didn't show up because he had just undergone eye surgery and was not allowed to fly. So I gave a very general talk about the ribosome problem and hinted that we were making progress. I left early to go back to working on the papers, so I missed Steve Harrison, who had arrived late for the meeting. But he met Richard Henderson, who told him we had essentially solved the structure – something I'd never told anyone there myself!

Steve had become quite exasperated about the ribosome race and wrote to me soon afterwards to explain his feelings. He was furious that Ada's latest manuscript had been rejected by *Nature*, apparently because the reviewers had told the journal that it was not a sufficient advance over our paper of the previous year. Steve, who had seen Ada's manuscript, vehemently disagreed. He said Richard had told him about our new structure, and he felt that if Ada had been able to publish a paper that was an improvement over our previous one, and our forthcoming paper was an improvement over hers, we would be leapfrogging over each other and both of us would get credit. At the time, I was not at all convinced. I felt that Ada was trying to prematurely scoop both Yale and us in her haste to publish an unfinished structure, and it was no wonder that the reviewers at *Nature* thought this was not appropriate. In the end, it was Steve who turned out to be right.

Ever since we had returned from Argonne with good maps, we had been engaged in a frenzy of activity. We had to make sure the structure was as complete and accurate as we could make it, while simultaneously writing up our results. Writing the papers would have been hard even if we had not been under the gun. The structure was far more complicated than anything we had solved before, so even thinking about how to describe it and highlight important aspects in a way that was both readable and understandable was a challenge. Occasionally, we would discard or rewrite entire sections, causing tempers to fray. Perhaps even more difficult than the text itself was making the illustrations to accompany it, a job largely carried out by Ditlev and Bil. Without careful pruning to show just the essential features, pictures of complicated molecules tend to look like a messy tangle of coloured pasta (because the chains of molecules are drawn as ribbons). The ribosome had so many components that even choosing colours that would make the components stand out from one another was a challenge. Ditlev, with his fondness for subtle pastel colours, would argue with Bil, who preferred strong and contrasting primary ones. Ditlev's computer skills were incredibly useful. Knowing that I would repeatedly ask for changes, he had written scripts for all the figures so that he could quickly make changes as necessary to make our points clear.

Looking back, the stress and adrenaline must have concentrated our minds because I am not sure we could have written anything much better even if we'd had more time. The papers were finished only a day before Brian was to leave for Santa Cruz and a few days before the Birmingham meeting. I was so worried about the manuscripts getting lost or delayed that I asked Bil to take the train to London and hand-deliver the four copies to *Nature*'s offices.

We were completely exhausted, but it was time to go public. Brian gave his talk on a Friday in Santa Cruz, and I gave the same one the following Monday in Birmingham. Brian was suffering from total exhaustion, and it must have showed, because his PhD adviser, Nacho Tinoco, unaware of Brian's furious pace in the previous months and the fatigue from travel and jet lag on top of it, chastised him for not appearing enthusiastic enough about the work. I felt sorry for Brian but knew that once the work was published, it would make his reputation. Brian wrote back saying that Ada's maps did not look as good as ours, but she had a model of the ribosome in which about fifteen of the twenty proteins were placed, as well as much of the RNA.

The lecture hall in Birmingham was in the basement of the conference centre in a long room in which the screen for the slide projector seemed like the size of a postage stamp, but we didn't care. Oddly, Ada and Marin van Heel, the electron microscopist, had a much bigger screen because they were ahead of the times – they were the only ones who used a computer to project their images while the rest of us were still using conventional slides in a carousel. In Birmingham, just a few days after her colleague Francois Franceschi had spoken in Santa Cruz, Ada claimed that she now had much better maps and had placed all but one of the proteins in her model but could not show her latest slides because of computer problems. But the proteins she did show seemed incomplete because there was no hint of the long snake-like extensions of the proteins that both the Yale group and we had seen in our structures. These extensions penetrated into the core of each subunit. Bil was furious and wanted to get up and make a comment, and I had to restrain him. There was simply no point in picking a fight that would make us look churlish and petty.

Our euphoria died again in the summer when the papers began coming out. The first was a pair of long papers in *Science* from the Yale group. They had used the same hexammine compound Jamie Cate had used, which was not surprising in hindsight – they had been aware of the possibility ever since he first used it, since he did that work at Yale when he was Jennifer Doudna's student. Not only that, they had actually obtained the compound from Jamie himself. A technical curiosity was that they no longer had a serious crystallographic problem called twinning. In their structure the previous year, which had come out along with ours, just after Denmark, they had shown that their crystals were twinned and suggested this was what had hindered progress on the structure for a long time. Twinning is every crystallographer's nightmare. In its worst case, the physical crystal is a composite of two different crystal lattices that give rise to spots in exactly the same positions on the detector. If one doesn't realize that the spots do not arise from a single lattice, the analysis of the data can yield nonsensical results. Even when you know there is twinning, you have to estimate how much each component of the twin contributes to a given spot, which leads to additional errors in the analysis. But soon after Denmark, one member of the Yale group made a fortunate 'mistake,' which raised the salt concentration closer to the original condition in which the 50S subunits were crystallized, and the twinning problem simply went away. It is still not clear to me if Ada's original crystals were twinned or whether the twinning was caused by the lower salt concentration in the solution the Yale group had originally used to freeze their crystals.

In any case, the Yale structure was spectacular and caused huge excitement. Tom Cech, who had originally discovered that RNA could be an enzyme and carry out chemical reactions, had himself written the short essay introducing the paper. He ended by saying that although it was one beautiful frame, we needed the whole movie of how the ribosome worked. Although this was certainly true, it seemed a rather grudging way to end his piece.

Next came Ada's paper, published in *Cell*. It was a big improvement over what we had published the year before but was not nearly as complete and accurate as ours. The next three weeks seemed interminable. I was annoyed at *Nature* for taking so long to publish what they should have been grateful to have in their journal; day by day I became increasingly depressed that people would perceive our work and hers as equivalent – and that we were just a little too late.

I need not have worried. When our papers came out, people clearly recognized that they described the definitive and complete structure of the 30S subunit, and it became the one that people have used since. So exactly as Steve Harrison had wished, Ada and I ended up leap-frogging over each other once more. In the end, just as he had hoped, we both got credit, but it was a long journey before that happened.

CHAPTER 14

Seeing the New Continent

THE ATOMIC STRUCTURES OF THE subunits were amazing. It was like landing on a new continent and encountering completely new and different terrain. Instantly, some key things stood out. One was that the ancient core containing the important parts was almost entirely made of RNA. The structures of both subunits strongly suggested that the ribosome had come out of an earlier RNA world, exactly as Crick and others had suggested over thirty years earlier.

In fact, the proteins were almost entirely on the outside and nearly all on the back side of the subunits so that the interface, the surface between the two subunits that bound the tRNAs, was made almost entirely of RNA. The proteins had long, snake-like extensions that penetrated the core of the ribosome. These extensions had lots of positively charged amino acids that neutralized the repulsive force of the negative charges on the RNA to allow it to fold better. The 50S subunit was twice as large as the 30S subunit, and the RNA was even more complex and intricately folded.

To locate exactly where the peptide bond was being formed, the Yale group had soaked into the crystals a compound that mimics the two amino acids bound to the ends of tRNA in the act of being joined. The location of the compound in the crystal structure

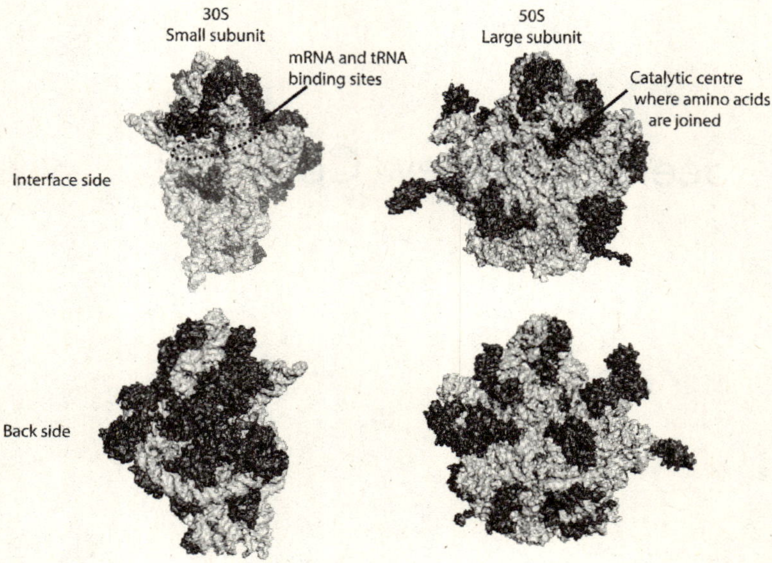

Figure 14.1 Front and back of the two subunits showing the darker proteins on a lighter RNA

identified the exact catalytic centre of the ribosome. It was surrounded entirely by RNA elements, showing that the ribosome was clearly a ribozyme, as had long been suspected.

The Yale group had also used biochemical data produced by their colleague Scott Strobel to produce a detailed mechanism for how the reaction of joining two amino acids might occur. But here they had overreached. The biochemical data did not reflect what was happening in the cell, and lots of chemists hit back at the proposed mechanism. That willingness to challenge is the great thing about science: no matter how important a discovery, people will attack any parts of it that they think are not right. In the end, a talented student of Tom's, Martin Schmeing, generated many more structures of different mimics of tRNA and amino acids bound to the large subunit. Using these structures as a guide, many biochemists, including Scott himself, figured out minute details of the reaction, such as how protons move from one group to another in the process. These experiments involved such sophisticated chemistry that I barely understood them. But it means we now know in greater detail than ever before how nature carries out one of its most essential reactions: the synthesis of a protein chain.

If the central task of the 50S subunit was to catalyse the joining together of amino acids to form a protein chain, the corresponding job of the 30S subunit was to make sure the genetic code on mRNA was read and translated accurately. Each codon on mRNA is 'read' by an incoming tRNA that brings a new amino acid to the ribosome. This recognition of the codon by the right incoming tRNA is called decoding, and the region surrounding it is called the decoding centre. Now that we had a structure of the 30S subunit, we thought we might see how mRNA and tRNAs bound to it to solve a long-standing puzzle in reading the code. The puzzle comes from the fact that there are sixty-four possible codons, of which three are used as stop codons, but there are only twenty amino acids, which means lots of codons code for the same amino acid. Because there are also fewer tRNAs than there are codons, many of them must read more than one codon. Frequently, but not always, the multiple codons that code for a particular amino acid differ only at the third position. When the genetic code was being determined, Crick noticed this and said that perhaps the tRNA could wobble a little and be more tolerant at the third position.

In other words, a match between tRNA and the codon would require exact base pairing at the first two positions but not always at the third. Why did the ribosome only accept tRNAs that paired perfectly at the first two positions but was more tolerant about mismatches at the third position?

The bonds that the partners make in a correct base pair (e.g., A-U or C-G) are stronger than in a mismatched one like U-G or A-C. But not by much, and that difference in bonding energy wouldn't account for why the ribosome was so selective about the right tRNA for the codon. The ribosome typically has an error rate of less than one in a thousand, or far better than any peptide synthesizer in the lab. It also does it amazingly fast, building about

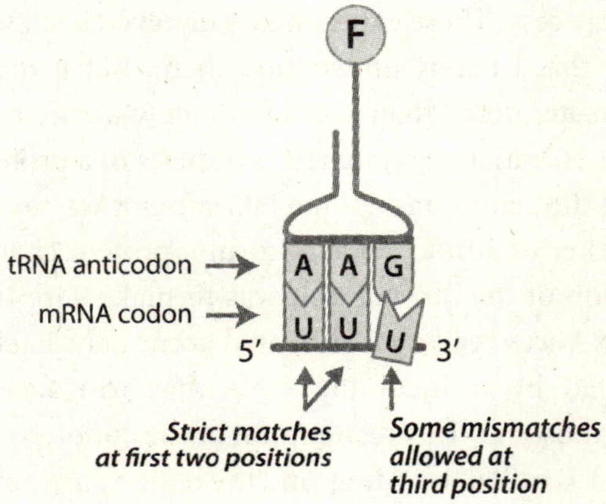

Figure 14.2 A tRNA has to pair perfectly with the codon at the first two bases but not the third, called the wobble base

twenty amino acids per second in a typical bacterial cell. So how was the ribosome so accurate? How did it know to reject tRNAs that were slightly wrong?

We already had a hint of what was happening from one of the antibiotics we had studied, paromomycin. The antibiotic was known to increase the error rate of the ribosome by causing it to misread codons. Our structure of the 30S with paromomycin showed that its binding resulted in two bases being flipped out of their long helix towards where the mRNA codon and tRNA anticodon would be. This suggested that these bases were sensing the groove between the mRNA and tRNA bases and were somehow stabilizing them, so that even an incorrect tRNA could be accepted. But the details of how this would actually happen were not at all clear because we didn't have mRNA and tRNAs in our crystal.

To understand how decoding might work, we would normally try to make a complex of mRNA and tRNAs with the 30S subunit (or better still with the whole ribosome, as we did later) to figure out what was going on. But it would be impossible with this particular crystal form for a peculiar reason. The site for the tRNA that holds the growing protein chain is called the P site and the one with the decoding centre that binds the tRNA bringing the new amino acid is called the A site. In these crystals of the 30S subunit, a piece of RNA from a 30S subunit called the spur was sticking into the P site of its neighbour. If we added mRNA and tRNAs to the 30S subunit before crystallization, the complex would block the contact with the spur from the neighbouring molecule and the crystal would never form. But there was another way around the problem.

As with typical proteins, ribosomal subunits in a crystal have water channels between them. This is how we could soak in small compounds like antibiotics into the crystal, and they would diffuse through these channels to find their target site on the 30S subunit. People had been soaking drugs and inhibitors into crystals of enzymes for a long time to understand how they worked.

However, we noticed that because the 30S subunit was so large, and the crystals were over 70 per cent solvent, the aqueous channels between neighbouring 30S molecules in the crystal were large enough that they could probably accommodate small proteins or RNA molecules that had over a thousand atoms – much larger than antibiotics. One of these channels led directly to the decoding centre. So could we soak an entire protein or RNA molecule into these crystals?

This had never been done before. Andrew Carter tested the idea with a protein factor called IF1 (for initiation factor 1) that helps the ribosome start translating, which was known to bind to the A site. He took 30S crystals soaked in IF1 to Grenoble and, on his return, triumphantly showed maps where he could clearly see the protein. So, soon after the 30S structure was solved, we now also had a snapshot of it with a protein that helped the ribosome get started.

This immediately started us thinking about doing the same sort of experiment with pieces of RNA that mimicked the codon on mRNA and a hairpin 'stem-loop' that mimicked the anticodon arm of the tRNA. The hope was that one piece would slide into the slot where mRNA normally binds and the other would interact with it like the anticodon arm of the tRNA. It seemed a little crazy but plausible.

The task of doing this experiment was given to James Ogle, a new graduate student. A German whose parents were English, he

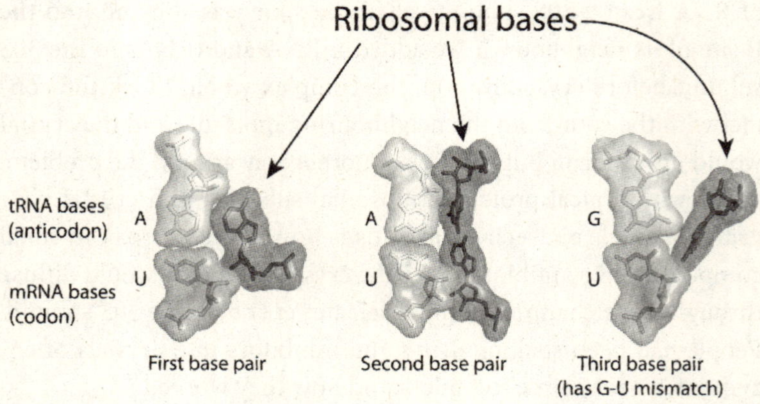

Figure 14.3 The ribosome recognizes the shape of the codon-anticodon base pairs at the first two positions but not the third

typified the new European who was multilingual and comfortable in lots of countries. He was bright and supremely self-confident and had lots of other interests, like being a gifted amateur violinist. James carried out the experiments and then went with Ditlev and the others to Argonne.

As soon as I got the data back from them over the Internet, I looked at the maps and we could clearly see what was happening. In addition to the two bases that had been flipped out by the antibiotic paromomycin, a third base had also changed. These three bases were indeed a reading head, as we had proposed earlier, and had inserted themselves into the groove between the codon and anticodon bases at the first two base pairs. In doing so, they were recognizing the shape of the base pairs between the tRNA and the mRNA at the first two positions but not at the third. In fact, one consequence of this was the whole subunit had closed in around the codon and anticodon in a way that would not have been possible if the base pairs at the first two positions had the wrong shape.

As Watson had first noticed when he and Crick were figuring out the double helical structure of DNA, the base pairs AT and GC (and the reversed TA and CG) all have almost the same shape, so a DNA helix could have almost any combination of base pairs and preserve roughly the same overall structure. The same is true for RNA, with a U instead of a T. So correct base pairs have a characteristic shape that distinguishes them from mismatches, and that is how the ribosome was discriminating between them.

Accuracy is a very important concept in biology, and the cell has evolved to have just the right amount. Just as with typing, there is always a compromise between speed and accuracy. Too much accuracy, and the process is too slow to sustain life. Too little, and too much of the wrong product gets made, with harmful effects on the cell. Some of the antibiotics like paromomycin lower the accuracy of the ribosome. So we had figured out the underlying structural reason for why the ribosome is so accurate and why the genetic code had this strange property of being a three-letter code but generally requiring perfect matches only at the first two positions. Just as for peptide-bond formation, it was all done by RNA, again supporting the idea that the ribosome came from an earlier RNA world.

This was all very exciting, but we had already hinted at what was happening in our earlier papers on the 30S structure with paromomycin, and I was worried because, in the meantime, Harry's team had improved their 70S structure from nearly 8 Å resolution the year before to 5.5 Å resolution. This was the same resolution that we had for the 30S in 1999. It was good enough to model in things that were known but not to build a new structure from scratch. But by then they didn't have to because they now had the atomic structures of both subunits as a guide to building their structure. In fact, for the 30S subunit, it was even the same species, so all they had to do was slot our 30S structure into their maps – the 30S part of their final structure was essentially identical to what we had published.

I was worried that even if they couldn't see the details directly, they might figure out what was happening, because they had the tRNAs and mRNA in place, and we had already alerted them to what might be happening when we described what paromomycin does the year before. So as soon as I could see the result, I began writing a first draft of the paper even before James and the others had returned from Chicago. I knew Harry's paper was due to come out in *Science*, so I contacted the editor and said this would be a nice companion piece that addressed a major puzzle in how the genetic code is read accurately. Luckily, they agreed, and our paper came out back-to-back with the 70S structure from Harry's lab, on which Marat was the first author.

A bonus to come out of the ribosomal subunit structures was their complexes with antibiotics. In addition to the three antibiotics we had originally located in the 30S subunit, Ditlev led an effort to determine the structure of three more antibiotics, including the clinically important tetracycline.

They all bound exclusively to RNA, just as Harry's work had predicted. They were the first published reports of antibiotics bound to the ribosome and helped us understand how they would stop the

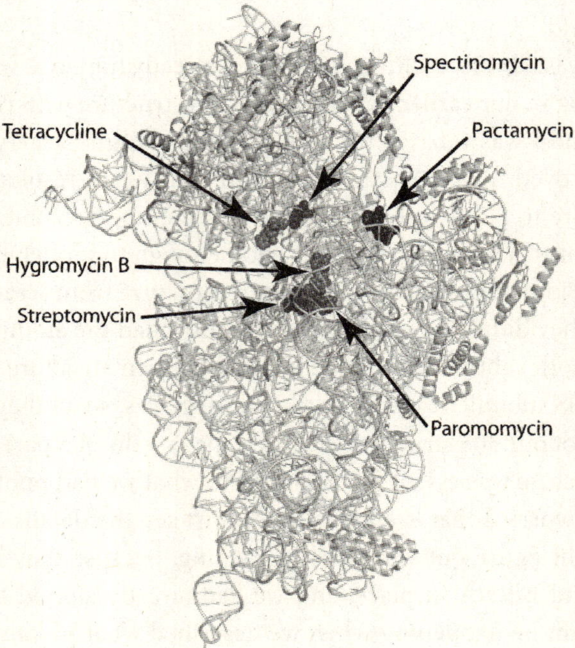

Figure 14.4 Antibiotics in the 30S subunit

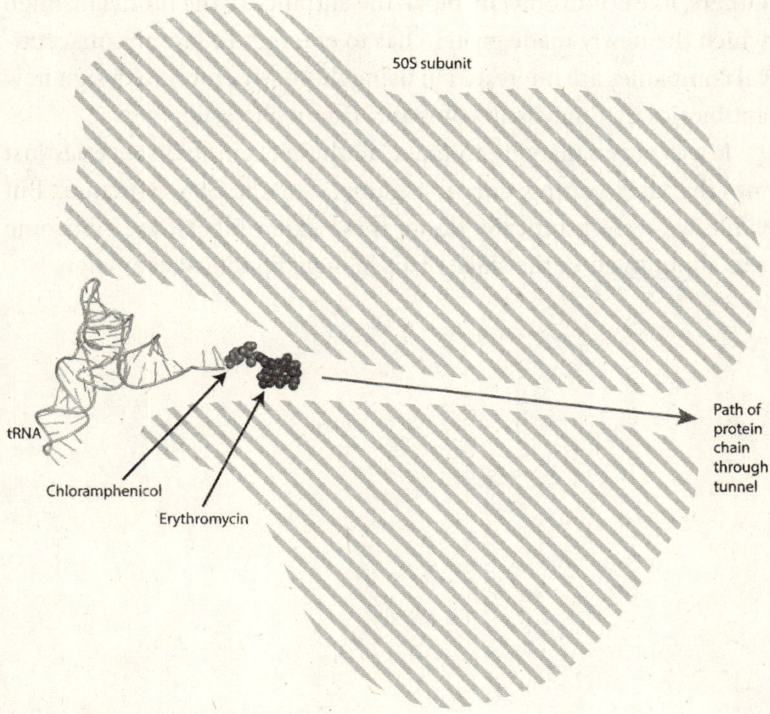

Figure 14.5 Two antibiotics in the 50S subunit

ribosome from working. One of them, spectinomycin, bound to a hinge point between the head and neck of the subunit. The head wobbles during the movement of the ribosome, and perhaps by locking it, the antibiotic prevents the ribosome from moving along mRNA. Another, tetracycline, works by preventing the new tRNA from binding, so amino acids cannot be added to the growing protein chain and the ribosome effectively stops dead in its tracks.

In the following months and years, both Ada and Tom published structures of many antibiotics bound to the 50S subunit. Some of them, like chloramphenicol, prevent the new amino acid from binding in the 50S subunit, so the protein chain cannot grow.

Others, like erythromycin, block the entrance to the tunnel through which the newly made protein has to emerge. Today, pharmaceutical companies are interested in using these structures to design new antibiotics that may help combat resistant bacteria.

It was very satisfying to make all these exciting discoveries just months after getting our first glimpse of the 30S structure. But while we were frantically doing these experiments, the ribosome was about to enter the stage of a prolonged political campaign.

CHAPTER 15

The Politics of Recognition

AFTER LONG BEING SEEN AS passé, the ribosome had developed a buzz ever since the initial breakthroughs towards its structure. There was talk of the ribosome being a candidate for big prizes, even a Nobel Prize, and people would bring it up over the next decade wherever I went. Even right after the initial progress reports on the structure in Denmark the year before, the ribosome structure groups were receiving lots of invitations to speak. All that year, I was more worried about being able to finish the structure and figured there would be plenty of time to talk about it later. But I can't pretend that the prospect of a major award didn't affect me.

Virtually every young student of science has had a fantasy about winning a Nobel Prize. It has become deeply embedded in popular culture, representing not only having done something great but actually being great. But as we mature, these fantasies quickly take a backseat to reality. Only a small fraction of scientists even come into personal contact with Nobel laureates; for everyone else, they take on an almost mythical quality and are not relevant to their own lives. So nobody goes into an area of research with the idea that there will be a big award at the end. Rather, we enter a field out of curiosity and interest in the problem, its applications, perhaps its benefit to the world, and, more pragmatically, job prospects.

Nevertheless, scientists are only human. Like everyone else, we can be ambitious and competitive and crave recognition. Instead of inculcating a feeling that the work is its own reward, the scientific establishment feeds this desire to feel special and somehow better than our peers at virtually every stage of the process. The corruption starts early, with small prizes throughout our education, then prestigious fellowships, then early career under-forty prizes. Later on, scientists hanker to be elected to their country's academies and then win even grander prizes. It is the darker side of a natural human desire to feel respected by our colleagues. However, all these prizes, awarded at different career stages, affect only a tiny fraction of scientists. Most of them go to people who work in elite institutions, have had powerful mentors and networks, and are on the fast track to fame and glory.

At the pinnacle is the Nobel Prize, but it is rare for someone to suddenly be awarded the Nobel without some hints that he or she is a serious candidate. As soon as someone does something considered significant, he or she is in contention for lots of prizes that are little known to the public. One might think that these prizes are all independent and, with the explosion of science, could be used to recognize many different scientists and discoveries.

In fact, the whole system is beset by a kind of cronyism. The various prizes often go to the same people, who are often well-known and powerful scientists. Often one bold committee makes the first award in a new field, and then other prize committees play it safe by following suit. This can quickly have a snowballing effect, with the result that the same luminaries pick up lots of awards. Moreover, the primary motive for many of these prizes becomes honouring the prize itself along with its nominating committee – rather than honouring the recipient or selecting good role models or highlighting interesting work in unfashionable areas. So instead of differentiating themselves from the Nobel Prize and complementing it, many of these committees measure their success by how many of their awardees go on to win the Nobel Prize and proudly advertise this fact. You could consider these awards 'predictors,' in the way that the BAFTA or the Golden Globe is often considered a predictor for the Oscars. In the worst kind of subservience, these prizes will not consider even a subject, let alone a person, that has already been awarded the Nobel.

How did the Nobel acquire its standing? It was instituted almost accidentally by a Swedish chemist, Alfred Nobel, who invented dynamite and parlayed it into a huge industry. Worried about his legacy, he decided that the bulk of his enormous fortune should go towards a set of prizes. Three prizes in the sciences – physics, chemistry, and physiology or medicine – were to be administered in Sweden, along with a prize in literature. A separate peace prize was to be administered in Norway. Curiously, there was no prize for mathematics.

The timing of the first Nobel Prizes in 1901 was particularly propitious. Its inception coincided with the kind of revolution in science that happens only once every few centuries. Physics had discovered quantum mechanics, subatomic particles, and relativity, changing our view of matter forever. These in turn revolutionized our understanding of the forces that hold molecules together and the mechanisms of chemical reactions, making chemistry a modern discipline. The discovery of genes and the inner structures of the cell revolutionized biology. Many of the earlier recipients of the Nobel Prize, like Planck, Einstein, Curie, Dirac, Rutherford, and Morgan, were giants who will be remembered forever. When combined with the staggering sum of money involved, which at the time was enough to guarantee the recipient financial security for the rest of his or her life, the Nobel soon became synonymous with greatness. It was certainly not infallible, making some terrible omissions, like Mendeleev, who discovered the periodic table, which is the basis of modern chemistry; Lise Meitner, who provided an explanation for nuclear fission; or Oswald Avery, who discovered that DNA is the genetic material. Occasionally there were also blunders, like the prize for lobotomy or giving one to

Johannes Fibiger for his now discredited finding that roundworms caused cancer, while simultaneously rejecting Yamagiwa Katsusaburo, who showed that chemicals in coal tar can cause cancer, which paved the way for the study of carcinogens.

If the science prizes are not perfect, other areas can be even more contentious. In literature, although some great writers have been recognized, too often the award typifies the preference of academic literature departments, going to obscure, unreadable writers or, even worse, to mediocre or bad ones. The prize has gone to long-forgotten writers – but not to Twain, Tolstoy, Joyce, Proust, Nabokov, Borges, or Graham Greene. In 2018, as I write this, the Swedish Academy that awards the literature prize is in disarray, with several members, including its first female permanent secretary, having resigned as a result of disagreements among factions, resulting in a decision not to award a literature prize this year. And what can one possibly say about a peace prize that was awarded to Arafat and Kissinger but not to Gandhi? Much later, in 1969, an economics prize 'in memory of Alfred Nobel' was instituted by the Bank of Sweden to cash in on the Nobel name. It has been awarded to economists with quite different and, to an outsider, contradictory views of their field. A particularly amusing year was 2013 when Eugene Fama and Robert Shiller shared the prize, which would be a bit like Darwin and Lamarck sharing a prize for evolution.

The prize is also subject to some rather arbitrary rules. Nobel specified that the prize was to be for a discovery or invention made during the previous year, but because it often takes years or even decades for the importance of something to become clear, this rule was very quickly discarded. A second arbitrary rule, that the prize would be awarded to no more than three recipients, was formalized only in 1968. In a slavish imitation of the Nobel, the Lasker Foundation adopted the same rule in 1997 for its prize, known as America's Nobels. The Lasker jury has been chaired for many years by Joseph Goldstein, a Nobel Prize winner who discovered the fundamental biology behind statins and has thus helped prevent millions of heart attacks and strokes. As someone on statins myself, I am personally a beneficiary of his research. In a recent article in *Cell*, he justified the rule of three by giving the number an almost mystical quality and comparing it to art's three-panel triptych. I think it merely shows that years of chairing a jury can make the criterion seem so natural and reasonable that even someone of Goldstein's stature and intellect can justify it by this exercise in numerology.

In fact, the rule of three is inappropriate today. When the Nobel Prizes started in 1901, scientists worked in relative isolation and met only once every few years. By the time they announced their findings, there was no question who had discovered what, and it was rare that more than three people would have contributed to the exact same discovery. In today's world, the germ of an idea shared at a meeting quickly spreads throughout the world and lots of people contribute to its development. It is not always clear whether the original idea or some later contribution was the truly groundbreaking advance. In sport, there is a clear way to measure performance – the biggest score, the fastest time, the longest jump, or the highest vault. But in science, figuring out three people who made the real difference in a particular field becomes increasingly difficult and subjective, if not impossible. Also, the explosion of science in the last half century has meant that lots of important advances never get the prize and it has increasingly become a lottery.

Increasingly, the rule of three means that year after year, there are complaints about people who have been overlooked. Although many big advances, like the discovery of the Higgs boson or sequencing the human genome, are done by large collaborative teams, the science prizes, unlike the peace prize, are not awarded to institutions. And though the Nobel amount is large, there are now prizes that dwarf it in cash value. For these and other reasons, the Nobel may be in danger of losing its unique, exalted status.

A recent potential rival to the Nobels is the Breakthrough Prize. It is the brainchild of Yuri Milner, a physicist turned entrepreneur and venture capitalist billionaire. He decided to reward famous physicists in string theory who would never get a Nobel because the theory is, so far, not capable of being experimentally verified and is thus more akin to natural philosophy than science. In one stroke, he gave nine of them a prize of three million dollars *each.* Subsequently, he convinced fellow billionaires like Sergey Brin and Mark Zuckerberg to join in, so there are now Breakthrough Prizes in the life sciences and mathematics.

The first recipients seem to have been decided rather arbitrarily by Milner and his associates, probably after consultation with some famous scientists, so it was not surprising that they were generally well known or well connected. The rule that future prizes are decided by a vote of former prize winners may simply perpetuate current fashions in science and favour well-connected people. I told Milner that I did not think this was a great way to award prizes, and it was – to use an analogy from his world – the equivalent of awarding prizes by the number of 'Facebook likes.' His response was that prize winners would want to maintain the prestige of their prize and would therefore be quite stringent about the quality of people they voted for. But this prize is in danger of becoming yet another closed club of elitists voting for people conforming to their own image of science.

The Breakthrough Prize is worth about eight to ten times the cash value of a shared Nobel Prize and is awarded with great fanfare in California, with Hollywood celebrities taking part in the ceremony. There are several laudable characteristics that distinguish it from many other prizes. It has been awarded to institutions and teams, which better reflects the industrialized nature of much of modern science. It does not have the arbitrary rule of three and has even occasionally been awarded to people who were left out of the Nobels because of that rule. It also ignores another Nobel criterion, which is that theories should have been experimentally verified to qualify. As a result, the Breakthrough Prize has gone to some brilliant physicists who did not qualify for the Nobel, such as Stephen Hawking (who died as I was writing this book). But just like other prizes, the Breakthrough Prize has almost never been awarded to someone who already has a Nobel (a few lucky recipients have really hit the jackpot by getting both in the same year).

In any case, despite the large difference in the award amount, I doubt that most people would want to trade the Nobel for a Breakthrough Prize quite yet, although that could change. Despite its problems and various rivals, the Nobel, because of its history, its exclusivity, and perhaps especially its public perception, still sits at the pinnacle. There are also other reasons it continues to be respected. The committees take their time, seeking expert opinions, giving prospective candidates the once-over by inviting them to meetings in Sweden, and deliberating carefully before making their decision. Nobody has questioned their integrity, even if increasingly frequently people have complained about their decisions. Also, the Nobel committees seem not to be swayed by politics or popularity to the same extent that others are and have often awarded the prize to scientists who were relatively unknown, even in their own country. In some cases, their national academies, embarrassed at having overlooked them, scramble to elect them as members the following year. A notable example is the award of the physics prize very early on to Marie Curie, at a time when she had little recognition and there were hardly any women working in science. She also went on to be the first person of any gender to win two Nobel Prizes.

Because many of the early Nobel laureates were giants in their field, the idea has taken hold – especially among non-scientists – that Nobel laureates are geniuses. In fact, the prize is not awarded for being a great scientist but rather for making a groundbreaking discovery or invention. Some of them may be extraordinarily brilliant, but others are just good scientists who were persistent or happened to stumble onto a major finding. Being in the right place at the right time often helped enormously. What Malvolio said in Shakespeare's *Twelfth Night* equally applies to Nobels: some are born great, some achieve greatness, and some have greatness thrust upon them.

But the label of genius that goes with the Nobel means that scientists, if they reach the stage where there is even a slight chance of getting one, hanker after it. To get one is – at least in the public's perception – to join the pantheon of the greats. Some hanker after it so much that it changes their behaviour, and their writings and public appearances have all the hallmarks of a political campaign. It makes them deeply unhappy and frustrated when, year after year, they fail to get one, and is a disease that I call pre-Nobelitis.

After the prize, there is post-Nobelitis. Suddenly, scientists are thrust into the limelight and bask in the public adulation that goes with it. They are asked for their opinion on everything under the sun, regardless of their own expertise, and it soon goes to their head. Some of them are long past their prime, having made their big discoveries decades earlier, and the renewed attention means that they spend their time wandering around the world, pontificating about all sorts of things. They become what I call professional Nobel laureates. Some laureates escape the disease, either because they are still very active scientists who simply ignore the distractions and continue to focus on doing the kind of good science that got them the prize in the first place, or because they use the prestige to do good for science in various leadership roles. A great example of the latter is Harold Varmus, who won the prize for identifying genes that in some circumstances can transform a normal cell into a cancerous one, but then became the director of the National Institutes of Health in the US and a strong advocate for biomedical research.

Prizes are often touted as a good thing for science by increasing its visibility with the public and providing the public, especially young people, with good role models. Nacho Tinoco, the famous physical chemist who was Brian Wimberly's mentor, once told me he thought the Nobels were good for science because they fostered competition among top scientists and spurred them to do their best work. They may be good for science, but they are not so great for scientists. They distort their behaviour and exacerbate their competitive streak, creating a lot of unhappiness.

All cultures seem to want their heroes and role models, so maybe prizes are a reflection of some deep-seated aspect of human nature and are not going to go away. Their intrinsic unfairness may simply be another manifestation of the fact that life is unfair. So far, no scientist has voluntarily turned down a Nobel Prize (some like the German Gerhard Domagk were not allowed to receive it by the Nazi government of the time). For an individual scientist, the prospect of broad recognition and financial reward is far too much to turn down.

When I started working on the 30S subunit, my focus was almost entirely on figuring out how to get it done quickly enough not to lose out in the race. My career up to that point was entirely prize free, but now with lots of people talking about the ribosome as a potential candidate for the big prize, it was hard not to be affected. I started worrying about my relative contribution and whether I would be perceived as a Johnny-come-lately rather than a pioneer. So whatever reservations I had, I too found myself caught up in the politics of the ribosome over the next few years.

CHAPTER 16

The Ribosome Road Show

SOON AFTER THE ATOMIC STRUCTURES came out in the late summer of 2000, I continued to avoid the politics because we were still focusing on the important problem of how the 30S promotes accurate reading of the code. One of the few invitations I accepted that autumn was to give the Stetten Lecture at NIH along with Peter and Ada. Almost predictably, Ada spoke for so long that, when I began my talk as the last of the three speakers, the entire time for the symposium had been taken up. Luckily, our host was kind enough to let me finish my talk, so I didn't suffer the same fate as Paul Sigler in Seattle many years earlier.

Right after the talk at NIH, I headed for Cold Spring Harbor, where I had been asked to speak at the same crystallography course that I had once attended as a student. At the airport, I was waiting in line to check in when I spotted Jim Watson just ahead of me. I introduced myself, and we spent the plane ride to New York sitting next to each other and chatting about the progress on the ribosome and why it had taken so long. Suddenly, apropos of nothing, he said I should forget about 'the prize' because with the people from Yale and that guy in California and the Israeli woman (obviously Harry and Ada), there was no room for anyone else. Watson showed up at my talk the next day – apparently a rare event, according to the organizers of the course. Presumably he wanted to size me up.

I thought it was an odd remark for him to make so soon after the structures had come out. After all, there was a lot of work to be done to show how the ribosome really worked and only time would tell who had really made the important contributions. So although I found his comments vaguely disturbing, I put them aside, thinking it would be a long time before there would be any major awards for the ribosome.

I couldn't have been more mistaken. In late 2000, only a few months after the atomic structures of the two subunits had come out, the Rosenstiel Award given by Brandeis University was awarded to Harry, Peter, and Tom for showing that peptide bond formation on the ribosome is catalysed by RNA. Although this was a major discovery, the ribosome was important for much more than being a ribozyme and was not even the first example of that. After all, the polymerases involved in replicating DNA or copying it to mRNA are considered hugely important, even though they are proteins like most other enzymes. So I couldn't help feeling that the jury had simply decided on the three people they wanted to honour and then written a citation that would exclude the others.

When I told Richard Henderson (a former Rosenstiel awardee himself) about this, he said that regardless of any prize considerations, I should accept more invitations to meetings and talks to get our story known if only to get proper recognition for our work. The dissemination of science is through the ancient human art of storytelling. Scientists are often too busy to read papers outside their field (or these days, even in their own field), and it is only when they directly hear the people themselves that they actually learn about the work and who did what.

A few months later, I received an invitation to speak at Cold Spring Harbor Lab again. Each year, the lab organizes a major meeting on a topic that it thinks has reached some watershed moment. The books that come out of these symposia can almost be read as a chronology of landmark events in the life sciences. In 2001, it was to be on the ribosome. I was only too delighted to accept but then came a rather surprising follow-up invitation.

Each symposium has two special lectures that are about an hour long instead of the typical fifteen to twenty-five minutes of the standard ones. One of them is meant for the participants, and usually one of the leaders in the field gives a longer survey of his or her work. The other, held on a Sunday, is a public lecture at which both the scientists attending the meeting and members of the public around Cold Spring Harbor are present. I was surprised to be asked if in addition to the specialized lecture on my work, I would also give the public lecture, called the Dorcas Cummings lecture. The reason they chose me was that, for a brief period around that time, I was the only one of the various groups to have published the structures of antibiotics bound to the ribosome, and they thought it would make a nice story for the public.

When Ditlev, Andrew, James, and I landed at JFK, we were quite surprised and amused to find that Cold Spring Harbor had arranged for a stretch limousine to meet us. Tom and I gave our talks on our structures on the opening night. Ada spoke the next day, and to remind the audience of her contributions, Don Caspar, the chair of her session, pointed out how her pioneering crystallization efforts had paved the way for the later breakthroughs. Ada announced that she had a new 50S crystal form from a bacterium that was '*E. coli* compatible'. Her argument was that *Haloarcula marismortuii* were archaea, a domain that branched off from bacteria very early on and have characteristics intermediate between higher organisms – eukaryotes – and bacteria. In addition, it grew in high salt. In her opinion, they were not really appropriate for studying antibiotics against bacteria. It was rather ironic for her to argue against using the *Haloarcula* 50S crystals as the Yale group had done, considering she had discovered them in the first place. In any case, it led to her focusing on her new 50S crystals for the next several years. I was relieved that I wouldn't constantly have to be competing and arguing with her about the 30S subunit. I joked to Tom that Ada was his to deal with now and – paraphrasing Kipling – no longer 'the brown

Figure 16.1 Harry Noller and Alex Spirin at Cold Spring Harbor in 2001 *(courtesy of Cold Spring Harbor Laboratory)*

man's burden.' That turned out to be largely accurate: for the next decade or so, Ada and Tom would argue about details of the 50S subunit, especially the structures of antibiotics bound to it.

A couple of days later, Harry gave the first of the two special lectures, the one meant for the scientists at the meeting. He was introduced by a former undergraduate student of his, Winship Herr, who was then a senior scientist at Cold Spring Harbor lab. Herr gave Harry perhaps the most gushing introduction I have ever heard for anyone and ended by saying that he had never known Harry to wear anything other than hiking boots, and he was sure that even when he met the king of Sweden, he would be wearing them. There was an awkward silence before Harry got up to speak about the entire ribosome that he had now interpreted in molecular terms.

The talk was a tour de force that showed how the structure could make sense of decades of biochemical and genetic work on the ribosome and its interactions with mRNA and tRNAs, and Harry's deep familiarity with and understanding of the ribosome was apparent in virtually every slide. But an outsider or even a non-structural ribosome scientist hearing it would not have easily realized that because of its low resolution, the structure from Harry's lab was not built from scratch but derived from the atomic models of the two subunits that were determined by Yale and us. Even that structure was based on years of crystallization expertise imported from Russia through Marat and Gulnara and crystallographic expertise and phasing strategy brought to the lab by Jamie Cate. Since credit normally flows up to the lab head anyway, very soon after the Santa Cruz structure came out, most ribosome and RNA biologists, unaware of all the factors that went into determining that structure, simply gave Harry all the credit. It became 'Noller's 70S structure,' and Harry began to be credited for solving the entire ribosome with mRNA and tRNAs.

Later that summer, when I visited Marat in Strasbourg, where by then he had started his own lab, he expressed his frustration by saying, 'I must prove that I am not a camel,' implying he needed to show that he was an equal and not just some vehicle for carrying someone else to fame and glory. Some years later, he was to show that, indeed, he was not a camel.

Harry's talk was on a Friday, and I had until 5 p.m. on Sunday to fret about my public lecture. Giving a scientific lecture or a public one is hard enough, but to give a public lecture in front of all your fellow ribosome scientists is really intimidating. Simplify too much and your peers will criticize you for not being accurate, while being too careful and rigorous will mean losing the public. Compared to my scientific talk, it took about ten times as long to prepare the public lecture.

It was a beautiful Sunday afternoon, and I was nervously standing around at the reception in the patio outside Grace Auditorium, shaking hands with various people and making small talk. Suddenly out of the blue, my son Raman appeared. Still in full schmoozing mode, I stuck out my hand and said, 'Nice to see you!' He looked at my hand rather amusedly before I recovered and gave him my usual hug. I then noticed that my sister, Lalita, a microbiologist then at the University of Washington in Seattle, was with him. Apparently, she had been giving a talk about her own work in New York City and had colluded with Raman to come and hear me. There were also some friends and neighbours from my old home on Long Island. So I felt the added stress of not only being judged by scientists and random members of the public but also by family and friends. Finally, it was time to go in.

The talk was going to be introduced by the lab director, Bruce Stillman, but he had thrown his back out and was bedridden. So Jim Watson introduced me instead. It was a long, rambling introduction, which started off with his early days of ribosome research. As he began describing the ribosome and what it did, I was afraid he was going to give my entire talk. He went on to say that scientific generations are about ten to fifteen years apart and that Peter Moore was his graduate student and I was Peter's postdoc, so in some sense, I was his scientific grandson.

When I got up to speak, I felt like starting off by saying, 'Thanks, Grandpa!' but restrained myself. When I finished, I thought I had done a good job of explaining the problem of translating the genetic code from first principles all the way to how ribosomes work and what they looked like and how antibiotics block them. My satisfaction took an instant hit when a member of the audience asked me if antibiotics make ribosomes eat bacteria and kill them. I then realized that, without context, large and small are meaningless. The ribosome is enormous for an assembly of molecules but tiny compared to a bacterial cell, which has thousands of ribosomes inside it. But on the whole, the lecture was well received, and I have used an updated version of it whenever I have to give a talk for the general public. After the lecture, all the invited speakers at the meeting were invited to dinner at the homes of various wealthy patrons around Cold Spring Harbor – another tradition of the symposium. The lecture also had another great side effect for me: after having survived that trial by fire in front of virtually the entire community of ribosome scientists, I was no longer perceived as an upstart but began to be considered one of the leaders in the field. We all went back to Cambridge on a high note.

The ribosome road show was now in full swing. We got invited to meetings all over the world. Soon after the Cold Spring Harbor meeting, many of us met in Pushchino, the home of Spirin's institute, the Institute of Protein Research. I had learned Russian as a teenager and had been enamoured of the country ever since. Harry and I gave the two opening talks, and I met Maria Garber and others for the first time. Spirin gave a marathon three-hour talk summarizing his life's work, but despite the length, his dynamic manner kept us engaged and entertained. At the banquet, I sat with Harry, and it was the best time I have ever had with him. As the vodka flowed freely, Harry got more and more effusive, making hilariously unguarded comments about the ribosome and various people who worked on it. Even without the vodka, I found myself getting vicariously intoxicated. Harry ended the evening by wishing me good night with a big bear hug.

Next was a meeting closer to home – in Granada, Spain. It was September 2001, and on a free afternoon, all of us were touring the beautiful Alhambra palace, an Islamic site that at its height was tolerant of both Jews and Christians. We heard that two planes had crashed into the World Trade Center but did not realize the full significance of this until we returned to our hotel and saw the shocking sight on TV of the twin towers collapsing in what was clearly an act of terrorism. An RNA biologist said to me, 'You realize what this means, don't you?' I thought he was going to say something about the erosion of civil liberties, emergence of a police state, or perennial war. But what he actually said was, 'You're going to have to shave off your beard.' The beard, which I had grown as a postdoc two decades earlier, survived another two years, but flying while bearded and brown subjected me to so many 'random' additional checks at American airports that I eventually shaved it off.

I was now being invited to speak at RNA meetings. The field of RNA biology had exploded over the previous decade as new roles for RNA kept being discovered, and because RNA plays a central role in the ribosome, they wanted to hear about my work. But people were studying the ribosome long before the role of its RNA became recognized, so I usually joke that I am an accidental RNA biologist, much like the character in Molière who discovers he has been speaking prose all his life without realizing it. Given the way I landed in Peter's lab, I'm not just an accidental RNA biologist but also an accidental ribosome biologist.

Harry, however, was the darling of the RNA community because he had relentlessly focused on the RNA part of the ribosome for so many years. The RNA Society, in their annual meeting in 2003, which was held in Vienna, gave him their first lifetime achievement award. These meetings consist of very short talks of twelve minutes with three minutes for questions before moving on to the next speaker. They are intended to give young scientists like postdocs or grad students a chance to talk about their work. Since Harry was getting a special award, he was allotted thirty minutes. They had asked me to chair the session on ribosomes where Harry would be the first speaker. He gave a very nice talk about the early experiments, which rather surprisingly suggested that the RNA in the ribosome might actually be doing something important and was not just a kind of scaffold on which to hang the various proteins, and how we could explain those results now that we knew the structure. So he, too, was an accidental RNA biologist, at least initially, and he pointed out how important it was to believe your own data and let it lead you where it may. He finished right on time, and I said I hoped the other speakers would follow the great example he had set.

But the next speaker was Ada, one of the few senior scientists to speak in one of the twelve-minute slots. At the twelve-minute mark, she seemed to have barely warmed up, and I indicated she should wrap things up. At fifteen, she kept going. At about twenty minutes, I stood up and asked her to stop, to no avail. By this time, some members of the audience started thumping their feet and doing a slow handclap. At around the twenty-five-minute mark, seeing my helplessness, the audiovisual team in the back cut off the projector and microphone. Ada didn't realize it right away because she was looking at her own laptop screen, but as soon as she noticed what had happened, she looked at me and asked if she could at least show the final slide crediting the people who did the work. When I asked them to turn on the projector, she had to skip over another ten or twenty slides to get to the end.

Harry was in fine form at the dinner afterwards. I asked him if in a billion years the ribosome would become a protein enzyme like the polymerases that duplicate DNA or copy it into RNA. I pointed out that, even today, the proteins on the ribosome had long tails that protruded into the RNA core, so perhaps we were watching the process of proteins slowly taking over. He laughed and said it would be like those cyborgs that take over someone's brain.

A number of invitations began arriving from Sweden, many of which were quite openly sponsored by the Nobel Committee for Chemistry. There was one on the ribosome at the Swedish Academy of Sciences in Stockholm and another on RNA biology on an island in the Sandhamn archipelago, just off the coast near Stockholm. Many of the usual suspects, like Tom and Ada, were also at these meetings. It was clear we were being auditioned.

The last of these that I attended was in Tällberg, the same picturesque village where Anders Liljas had organized a ribosome meeting before all the breakthroughs, when I was still in Utah. This time, in October 2004, it was on the entire central dogma – the maintenance and flow of genetic information from DNA to RNA to eventually make proteins. There were talks on DNA replication, on transcription of DNA into RNA, and, of course, on the ribosome. In addition, there were people who worked on detecting proteins in cells and imaging cells, as well as on telomeres, the ends of chromosomes. A lot of very famous names were there, including Bob Roeder, who had discovered that higher organisms have three types of RNA polymerases to make three types of RNA; Roger Kornberg, who had not only discovered the nucleosome, the fundamental unit of how DNA is packaged in the cells of higher organisms, but had also gone on to solve the structure of an RNA polymerase from a higher organism; Elizabeth Blackburn, who had discovered the enzyme telomerase that maintains the ends of our chromosomes; Roger Tsien, who had figured out how to make proteins glow with different colours so that they could be used to label various structures in the cell; and many others.

When the programme arrived, I noticed that the ribosome session was on the third day, but the opening evening had Aaron Klug and Harry as the only speakers. Adding two and two to make five, I assumed that this was a sign that the Swedes had already anointed Harry for the prize by pairing him with Aaron, a long-standing laureate. But it turned out that they had merely accommodated Harry, who had to return to Santa Cruz early to teach his course.

Harry explained how removing the tails of proteins that protruded into the tRNA binding site did not affect the ribosome's ability to function and tied it to his very early observations that suggested that ribosomal RNA was involved in tRNA binding. Although I was sitting in the front row, it was not at all obvious from Harry's talk that these tails and many other details he showed had originally come from our atomic structure of the 30S subunit. I was pretty furious and put it down to the politics of the ribosome. Later, someone asked me how he could have figured out all those details at 5.5 Å resolution and I snapped, 'He didn't.'

I sat in the back with John Kuriyan, a fellow Indian American who is a professor at Berkeley and one of the smartest people I know. Since we both saw this was something of a beauty contest, we decided to have some fun by giving people a score as they do in the Olympics for gymnastics, so after each talk we'd say '8.0!,' '5.0!,' or occasionally '9.9!' It was astonishing how closely our scores agreed.

John himself gave a spectacular talk about a protein that wraps around DNA during its replication, which involves a large elaborate machinery. At the time, he just had the structure of the protein alone, without DNA, but made a very plausible case for how it would bind DNA. Some of the repeating modules in the structure were in exactly the right positions to track along the grooves of the DNA helix.

Afterwards, at lunch, I was sitting across from a member of the Nobel committee, who was flanked by two distinguished 'contenders.' The committee member said he was not too convinced by Kuriyan's talk. The contenders nodded in vigorous agreement. I begged to differ and said I thought John had made a plausible case, and his lab was probably testing his ideas as we spoke. The committee member thought about this for a few moments, changed his mind, and agreed with me, whereupon the contenders instantly changed their minds along with him. Considering their scientific stature, I thought this sycophantic attitude was a little pathetic but not entirely surprising. Some of them, realizing the nature of the meeting, were extremely nervous, almost like graduate students before their thesis exam, and one of them was actually hyperventilating before his talk.

Aaron Klug introduced the session on transcription – the copying of DNA into RNA. He gave a rather pointed introduction, saying that eukaryotic transcription was much more interesting because in higher organisms transcription was highly regulated, which made it more complicated and important, and structures were key to understanding transcription. It was almost as if he was singling out Roger Kornberg, who like so many others in this book had also spent time at the LMB as a postdoc, where he had discovered the nucleosome. Ever since, Aaron had a high regard for Roger, whom he considered a protégé. Roger certainly justified Aaron's opinion by giving one of the best talks of the meeting. Unlike many scientists who are either good biochemists, geneticists, or structural biologists, Roger was superb in every aspect. He also gave beautifully constructed and illustrated talks, speaking in complete, coherent, flowing sentences without the ums and errs that characterize most speakers.

Tom gave two talks at the meeting because he worked on DNA and RNA polymerases as well as the ribosome. The ribosome session was after lunch. I was happily chatting away when I was surprised to see that people were starting to go in. I thought we had plenty of time left but suddenly realized that my watch must have stopped. Tom said he was wondering why I looked so relaxed. The session went well, and everyone, including Ada, was very well behaved and kept to their allotted time. Ada spoke about one of her more interesting discoveries – that a large region of ribosomal RNA around the peptidyl transferase centre was symmetric so if you rotated half of it by 180 degrees, it would superimpose on the other half. This suggested that the catalytic centre of the ribosome may have originally arisen by gene duplication, giving rise to an RNA fragment twice as large, but now with a twofold axis of symmetry.

By the time of the meeting, we had begun relating our structural work to some elegant experiments that Marina Rodnina, now a Max Planck director in Göttingen, had done on the rates of the various steps in accepting a new tRNA into the ribosome and how the rates differed for an incorrect tRNA. Our structural data agreed rather nicely with her interpretation of her own results, so I had made a point of this in my talk. After the session, Tom complimented me on my talk, and I joked that when you're Avis, you have to try harder.

Måns Ehrenberg, a well-known ribosome scientist from Uppsala in Sweden, was the only Swedish speaker at the symposium. He was a kind and thoughtful scientist but often had an earnest, almost dour appearance that reminded me of a character from a Bergman movie. Reflecting his desire to get right to the bottom of things, many of his papers were long, almost impenetrable tomes. Some of them reminded me of a Zen Buddhist koan, which was appropriate considering Måns's interest in Buddhism. Måns, who had done a lot of early studies on accuracy in the 1970s and 1980s, was peeved that neither Marina Rodnina nor I had mentioned his earlier studies and made a point of this before plunging into the main topic of his talk. His annoyance only increased when, at the end of his talk, Marina criticized its very premise.

So although we had been very friendly for several years – he had once invited me to give the annual Linné Lecture at Uppsala University – Måns came up to me at the banquet on the final evening and berated me for ignoring his work. I finally got annoyed and started arguing with him, when his colleague came over to defuse the situation.

Soon after we returned from the meeting, Tom Cech, the scientist who had first shown that RNA can catalyse reactions and was at the meeting, wrote up a summary. Cech always seemed more interested in the fact that the ribosome is a ribozyme than in any other aspect of the problem and mentioned our work only in passing. Perhaps his opinion about what was really important and interesting was generally shared. To cap it all, I later learned that Måns had joined the Nobel Committee for Chemistry.

Even though Måns and I stayed on friendly terms, given our argument at the banquet and his subsequent membership on the committee, I slowly became reconciled to the fact that I would never be a serious contender for the prize. Although disappointing, it was a relief in some ways. After having been blithely just doing science for most of my life, I had found the politics of the previous few years uncomfortable and distracting. I turned down nearly all subsequent invitations to meetings in Sweden in the next few years. Now I could go back to focusing on the work. It was time to help make the ribosome movie.

CHAPTER 17

The Movie Emerges

THE FIELD HAD REACHED THE limits of what we could learn from the structures of the two separate subunits of the ribosome. Some very localized functions, like the reaction that forms the peptide bond or how tRNA interacts with the codon, could be studied in individual subunits. So could the binding of antibiotics. But now we needed to understand how the ribosome selected tRNAs, moved along the mRNA, and started and terminated.

In many ways, understanding the process was like understanding any other machine. To understand a four-stroke internal combustion engine, you need to understand the various stages of taking in a mixture of fuel and air, compressing it, and having a spark ignite the mixture to create the power to push the piston, which can then rotate a crank and eventually move the wheels. You need a movie.

The ribosome, too, had its cycle, and we wanted to collect as many possible snapshots at different states of the process.

We already had fuzzy images of many of these states from electron microscopy. Largely led by Joachim Frank's lab, these images had given us our first look at different functional states of the ribosome. But although the resolution of the technique had improved

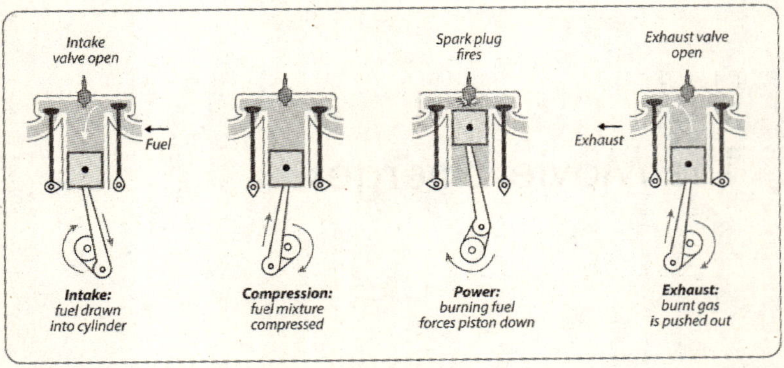

Figure 17.1 The cycle of a four-stroke internal combustion engine

gradually over the years, it still wasn't able to get even close to seeing the details of the chemical interactions, and how the machine worked in molecular terms remained mysterious. You could think of them as images shot out of focus so that the details are not clear. There seemed no alternative to the hard grind of trapping the ribosome in as many states as possible and laboriously trying to crystallize them, a process that could take years of effort for each state with no guarantee of success.

At that point, there wasn't even a detailed atomic structure of the whole ribosome. The structure from Harry's lab was at low resolution and moreover did not actually represent any particular step of the process. So a new race began, but unlike the previous head-to-head two-way races, this race involved virtually everyone who had done ribosome crystallography. In addition to some of the original groups, like Tom and Peter, and Harry, some of the people in Harry's lab, like Jamie Cate and Marat and Gulnara Yusupov, now had their own labs, and all of us were trying to get a high-resolution structure of the whole ribosome. At a meeting in Paris, Tom and I were talking to someone who thought the ribosome was 'done,' and we quickly corrected his misapprehension. He asked what would happen, and Tom said that some day someone would show up at a meeting with a big smile on his face, much to the disappointment of the others.

By then, the people who had worked on the 30S subunit in my lab had gradually moved on. The first of the new team to arrive was Frank Murphy, who overlapped initially with many of the 30S people. He had come from Denver and had a pleasant manner but combined it with a deadly sarcasm that was always delivered with a smile. Like Ditlev Brodersen, he combined great computational and crystallographic skills with excellent biochemical skills, and like Ditlev, he was also very systematic and organized. He was also a great teacher who helped train a lot of people in the lab and took charge of many of our synchrotron trips.

We thought it would be a good idea to start by reproducing the 70S crystals that the Yusupovs had reported from Harry's lab, just as we had used the Russian conditions as a starting point for the 30S subunit. Although we had some hints from Maria Garber that a chromatographic column had been useful for the Russian effort, we couldn't reproduce the crystals. Something was missing in the reports, or we were doing something differently. Even the original Russian report on the 30S and 70S crystals didn't have any useful information – as I found out when I used my limited knowledge of Russian to look at it.

On one of my mammoth trips as part of the ribosome road show, Peter and I were sitting around a table in Buenos Aires with Marat. We told him neither of us could reproduce his crystals using his published procedure. Marat said it depended on getting little details right. When we asked what details, he replied that he couldn't really tell us since some of them were only known to Gulnara. At this, Peter burst out laughing, saying, 'I guess this means neither of you will be able to spill the beans if you're captured!' In fact, I heard that Marat and Gulnara had kept their notes in Russian, and when they left Harry's lab, it seemed that he couldn't reproduce the crystals for a while either. Eventually, Harry got back on track when he hired Sergei Trakhanov, another key member of the Russian group that had originally crystallized the ribosome.

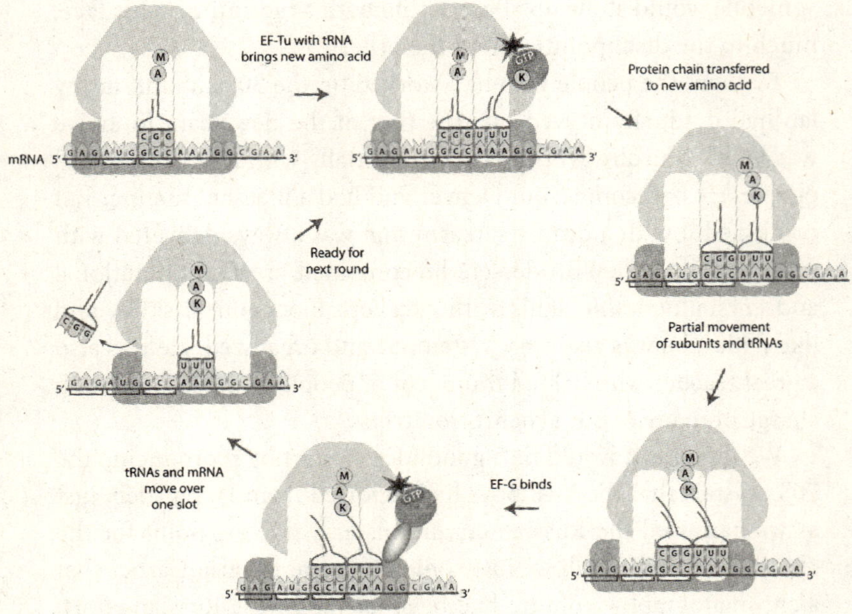

Figure 17.2 The ribosome elongation cycle

In any case, Frank kept plugging away at the problem with Mike Tarry, a technician who had replaced Rob, to figure out how to purify ribosomes that were good enough to crystallize. They learned how to purify away ribosomes from an enzyme that degrades RNA and were able to obtain some crystals that were in a different form and were about as good as the Santa Cruz ones – but would certainly not get us to atomic resolution.

With so many people going after a high-resolution structure of the whole ribosome, we thought it would be best if the lab focused on a few different states. It is never possible to predict which state will yield crystals, and all of them were interesting. This way, even if we got scooped on some of them, we wouldn't be completely out of the game. Frank decided to try to trap the ribosome right at the moment when a protein factor EF-Tu delivers the tRNA that brings the new amino acid to the ribosome. Our studies had already shown that the 30S subunit ensures that the tRNA is the correct one by monitoring the shape of two of the three base pairs between the codon and anticodon. Like many of the factors that bind to the ribosome, EF-Tu is called a GTPase because it cleaves off two of the three phosphates in GTP (guanosine triphosphate) and, in doing so, liberates energy. Because this hydrolysis of GTP is essentially irreversible, it is analogous to the spark and resulting combustion in an engine. For EF-Tu, the question was how this match between the tRNA and codon would result in the factor hydrolysing GTP quite far away from the codon. It is a bit like asking how turning a key that fits correctly in the ignition switch can fire up the engine a long distance away. The hydrolysis of GTP also acts like a switch that results in EF-Tu letting go of the tRNA. The end of the tRNA is then free to swing into the peptidyl transferase centre in the large subunit where the growing protein chain is transferred from the tRNA in the P site to the new amino acid on this tRNA in the A site.

While we were struggling to make progress for many years with *Thermus* ribosomes, we heard that Jamie Cate had crystallized the *E. coli* ribosome. Soon afterwards, we learned that he had crossed the magical threshold of 3.5 Å resolution, so he could begin building an atomic structure. This was a big breakthrough for two reasons. It would be the first high-resolution structure of the entire ribosome. It would also be from *E. coli*, the standard bacterium on which nearly all the biochemistry and genetics of the ribosome had been done. Ever since Ada had first obtained crystals of the large subunit from a thermophile, but only tiny microcrystals from *E. coli*, the general assumption was that, in order to get crystals, you needed extremophiles – species that grew in extreme conditions like very high temperatures or very high salt. Jamie was obviously young and bold enough to ignore conventional wisdom.

I was one of the referees of Jamie's paper in *Science*. Although I had initially felt a twinge of disappointment that it hadn't come from our lab, it was very exciting to see this major advance. Not only was it of the *E. coli* ribosome, but it was also the first time you could see the details of how the two subunits were held together, and it contained hints about how the ribosome would change as it moved tRNAs through it. There was one bit of bad luck for Jamie. His crystals were in a form in which the ribosomes couldn't bind mRNA and tRNA, so they were of limited use for getting snapshots of the ribosome at different steps of the process. This meant that most of the projects in our lab were still alive.

Anyway, my disappointment at having been scooped was short-lived because of two others who had joined the lab. The first of these was Maria Selmer. She had arrived from Anders Liljas's lab where she had solved a structure of a protein that splits the ribosome apart at the very end of translation so that it can start all over again. Like Ditlev, the other Scandinavian I've had in the lab, she was intelligent, organized, pleasant, and cheerful and just generally well rounded. It made me wonder if Scandinavians are doing something right in the way they bring up their children, or whether taking them from dark Scandinavia and placing them farther south, even if only as far south as Cambridge, made them particularly cheerful and free of any Bergman-like angst. Maria was noticeably taller than the males in the lab, so at the next annual Christmas skit, a tradition in which people satirize their colleagues and events of the past year, my group dressed up as Snow White and the seven dwarfs.

Maria was joined about a year later by Christine Dunham. I had met her at a meeting in Erice, Italy, and had been impressed by her intelligence and spark. At the time, she was a grad student with Bill Scott in Santa Cruz where she had acquired a real expertise in RNA chemistry and crystallography, so when she asked if she could work in my lab, I was only too happy to say yes. Of course, she also knew all of Harry's crowd from her time in Santa Cruz. She ignored my advice that it would be a waste of time to apply to the American Cancer Society for a fellowship and thus saved my lab a great deal of money when it was awarded. She was energetic and cheerful and shared my irreverent sense of humour and fondness for gossip.

Also around this time, Ann Kelley, who had been a technician at the LMB with others for almost twenty years, joined my group. She became a mainstay of the lab, producing ribosomes, tRNAs, factors, and anything else we needed so people could focus on their individual products. She was a feisty, red-haired northerner who would frequently get fed up with all the foreigners in the lab complaining about the local weather or customs.

Maria and Christine made a good team and decided to work on the problem of how tRNAs and mRNA move through the ribosome, a step referred to as translocation. Almost thirty years earlier, Mark Bretscher at the LMB had proposed that translocation occurs in two steps: the tRNAs first move in one subunit and then in the other. This provided a rationale for why there were two ribosomal subunits in all organisms. It took another twenty years before Harry and his student Danesh Moazed worked out methods that showed that indeed this was happening, and the ribosome entered an intermediate state where the tRNAs have moved relative to just the large subunit. In a second step, the tRNAs and mRNA move relative to the small subunit, completing the translocation. In 2000, Joachim Frank and Raj Agrawal showed that during this process, the two subunits rotate relative to one another. So the ribosome moved along the mRNA like a two-part machine ratcheting along a string. The second step of the movement was accelerated by another protein factor called elongation factor G, which also hydrolysed GTP during the process. Again, the details of how this happened and how EF-G facilitated the process were still unclear.

When Maria arrived in my lab, she was married with a young daughter and chose to have a very regular schedule, unlike the mostly single – or at least childless – members of the lab who would be there at all hours. I myself have always worked regular hours ever since I was a graduate student with two young children and felt the balance it gave my life made me both happier and a better scientist. About a year into her stay, she came into my office to tell me she was pregnant. I was happy for her and congratulated her, but deep down I was concerned that her maternity leave would disrupt our efforts in the middle of another intense competition between several labs. I was foolish to have worried, for Maria did more for the 70S phase of our work than almost anybody. It brought home to me in a very real way that someone can combine the jobs of parent and scientist and be very successful at both, provided they are in a supportive and flexible environment. In Maria's case, having a very sympathetic husband must also have helped enormously.

Later that year, Maria, now back from her maternity leave, had found a new condition that produced a different *Thermus* crystal form from anything previously known. The crystals turned out not to contain EF-G, even though the factor had been added to the ribosomes before crystallizing them. Still, they contained mRNA and tRNA and diffracted better than the ones from Harry's lab in 2001, and we figured some progress was better than none.

She and Christine set about systematically trying to improve the crystals and went with the others to the Swiss Light Source synchrotron in Villigen near Zürich. This time too, I got a surprising message from the team, but unlike the guillotine experiment years earlier, this was a pleasant shock. They had collected data to 2.8 Å resolution. This meant the maps were actually more detailed than even our 30S structures. However, there were two independent copies of the ribosome in the crystal – which meant building a million atoms into the structure.

It was an effort that reminded me of the old 30S days. In addition to Frank, Maria, and Christine, there were two graduate students who had joined the lab who were roped into the effort. Albert Weixlbaumer had done his undergraduate work in Vienna. He showed up for his interview with a number of body piercings, looking like the chief representative of a very hip Viennese counterculture. I accepted him because he did very well on his interview, but he then immediately asked if he could delay the start of grad school for several months so he could explore South America. I feared I had made a huge mistake, but once he arrived, there was no questioning his dedication. Albert was hardworking and was one of the most intellectually curious people I have ever had in my lab. By this point, he had become quite frustrated with his work on initiation complexes, a problem that scientists would make headway on only a decade or so later using a new technique.

The other student was Sabine Petry, who came highly recommended by Harald Schwalbe, a scientist in Frankfurt whose work I knew well. She was well over six feet tall and towered over everyone else in the lab, even Maria. Apart from her terrific undergraduate record, she had played on the women's basketball team for Germany. In many ways, she fitted the stereotype of the organized German. When she found out we didn't have regular group meetings, she was shocked. I told her if she wanted them so badly, she could start them. She did this right away and strictly enforced the roster of which lab member should speak each week. On another occasion, we came to the lab to find that all our old gel boxes had been thrown out and replaced with new ones coded with coloured tape that indicated what they could be used for; a stern notice above the boxes warned people not to mix them up.

The entire team had to work together for weeks in the graphics room to build this enormous new molecule. It wasn't quite as difficult as building the 30S and 50S subunits from scratch, but it was still a lot of work, especially since the 50S was a different species from anything that had been solved before. When it was finished, we could see in precise detail how the mRNAs and tRNAs interacted with the ribosome. It had been known for almost fifty years that lowering the magnesium concentration would make the subunits dissociate and raising it would cause them to come together again. Finally, we could see why: many of the contacts between the two subunits were through the positively charged magnesium ions, which mediated the contacts between the negative charges on the RNA phosphates. In other places, the magnesium ions allowed the RNA to fold tightly by neutralizing the negative phosphate groups and allowing them to come closer together.

As soon as we wrote up and submitted our work, Maria went to an RNA meeting in America while I went to Erice again. To our chagrin, Maria discovered at her meeting that Harry too now had an improved structure. In an echo of the 30S race six years earlier, the two structures were published nearly simultaneously. Harry's new structure was certainly at a higher resolution than the one they had in 2001, but just as with the two earlier 30S structures, ours became generally accepted as more accurate and complete.

Just as we had with the 30S structure, we were hoping to milk the new 70S crystal form to do interesting follow-up studies on functional questions but ran into a problem right away. When the lab tried to reproduce the 70S crystals that had given us such great diffraction, the results were highly variable. Some would diffract OK, but not nearly as well as the original crystal they had taken to Switzerland. Others wouldn't diffract at all. Even after we thought we had fixed that problem, we were having trouble getting anything to bind to the ribosome.

So the 70S crystals were not proving to be the goldmine that the 30S crystals were. Once the 70S structure was published, people started to leave the lab again. Frank went off to work at Argonne, where Malcolm Capel now had a senior position. Maria found a faculty job in Uppsala, and Christine found one at Emory University. Sabine had gone off to do a postdoc at UCSF. Of the old 70S crew, only Albert remained.

Around this time, the next ribosome meeting was to be held in Cape Cod in early June of 2007. It would be the first time I would talk about the high-resolution 70S structure after it was published. By this time, I was fed up with the politics of the ribosome, so I began my talk by being uncharacteristically blunt in my assessment of our 70S structure compared to the one from Harry's lab published around the same time, by pointing out some of the differences.

I also felt that going all the way back to our 30S structure, people had failed to appreciate how much our structures had been used not only to interpret biochemical data but also to correct other structures. To make a point of this, I suddenly showed a picture of an Ordnance Survey map of my village of Grantchester. I could feel the audience wondering what on earth this had to do with the ribosome. Perhaps they thought I had lost my mind. I went on to point out that the Ordnance Survey is thought to sprinkle a few spurious features in their maps so that if other map makers merely copied them rather than doing their own surveys, they would be found out. As scientists, I continued, we don't deliberately introduce bogus features, but we do make mistakes. I added that by using a new detector we had recently collected much higher-resolution data, which allowed us to correct some mistakes in our original 30S structure, and sarcastically said that it was odd how those exact same mistakes had appeared in other people's structures too. The next morning, Harry robustly defended his 70S structure, but was almost immediately rebutted by Tom, whose postdoc Miljan Simonovic had done a careful analysis of the two structures and come down unequivocally on our side.

Afterwards, Anders Liljas came to me and said he thought my behaviour was uncharacteristic. I replied that I was tired of people using our work without giving us credit. 'Why attack when you're ahead?' he asked. I didn't realize it at the time, but it was the first hint that my work was well regarded in Swedish circles. Still, I was happy to leave only a day later to briefly attend another meeting for a day before heading off to Northborough, Massachusetts, to watch my son Raman and his partner Melissa get married in the backyard of her father's house. It was a spectacular summer day and great to meet old friends and relatives while celebrating the young couple's happiness, and forget the problems and politics of science.

On my return, we were unable to make progress for several months. One day, Hong Jin, a new postdoc who had done her PhD in Peter's lab at Yale, walked into my office. She was new to crystallography and asked me how the procedure to freeze crystals worked. This process is very simple. You transferred the crystals into a cryoprotectant: a solution that had exactly the same composition as the original solution the crystal was grown in, but in addition, had some compound that was like an antifreeze, such as glycerol, alcohol, or ethylene glycol. She said that was not what we were doing. When I heard what we actually were doing, I suddenly realized why nothing else had been working with the new crystals.

In transferring the crystals to the cryoprotectant, my lab members had forgotten to include some of the original ingredients, and one of them was magnesium. Of course, all of us knew that magnesium was important not only to hold the subunits together but also to keep the RNA in the ribosome folded properly. No wonder the crystals were so variable in their quality. The crystals probably started off fine, but we were destroying their order by leaching out the magnesium before freezing them. It also explained why nothing we were trying, like binding release factors, was working. I kicked myself for not seeing this huge blunder that had cost us almost two years at this point. But at least we could now start making progress.

One major unsolved problem we had wanted to understand by using these 70S crystals was how the ribosome knows when to stop at the end of the gene sequence. The end of the coding sequence contains one of three stop codons (UAA, UAG, or UGA), which don't code for an amino acid but signal that the end has been reached. When one of these stop codons enters the A site of the ribosome, it is recognized by special protein factors called release factors, which bind to the ribosome and cleave off the newly made protein from the tRNA. In bacteria, there are two of them, called RF1 and RF2. How release factors recognized stop codons and how they cleaved off the protein were fundamental questions – after all, they had to do with how the ribosome knew when to stop at the end of a gene sequence and release the newly made protein product. Sabine Petry had been able to bind release factors to ribosomes in our earlier low-resolution crystals. We could never get them to bind to the new high-resolution crystals, but now we knew why.

It was almost too late. Soon after we discovered the magnesium problem, Sabine, who was now a postdoc at UCSF, wrote to say that Harry's lab had reported in a Bay Area talk that they had solved the structure of RF1 bound to the ribosome. The ultimate irony was that Harry's lab had abandoned their own crystal form, which he had vigorously defended in Cape Cod, and simply reproduced ours but without making the same blunder we had when we froze the crystals. I suppose it was actually the ultimate compliment, but it still hurt.

Harry's lab had solved the structure of the ribosome with RF1, the release factor that recognizes the stop codons UAA and UAG. So we started a mad scramble to do the same with RF2, which recognizes UAA and UGA. The idea was that between the two factors, we would know how the two release factors distinguished the three stop codons from the other sixty-one codons that code for amino acids and are recognized by tRNAs.

Albert, who had stayed on and had the most experience of doing 70S crystallography, led the effort, closely helped by Hong, who had first alerted me about the problem. As on previous occasions, the entire lab chipped in. This time they were helped by two new graduate students, Caj Neubauer and Rebecca Voorhees.

Caj started to work on a problem slightly peripheral to the main translation pathway, which is how ribosomes are rescued by a special molecule when they get stuck on a defective mRNA, for example one that lacks a stop codon and cannot terminate properly. He was a quiet and earnest young man who had done his undergraduate work in Germany. Caj was also a diplomat whom everyone liked. He would often mediate when factions broke out in the lab.

Rebecca came from Yale, where her undergrad adviser Scott Strobel (who had done so much work on how the peptide bond is formed by the ribosome) told me that she made better comments on his grant application than even many of his postdocs. Unusually, she initially came to do a one-year master's programme before going on to medical school back in the US, and even more unusually, she had brought her own project with her, which was to make two tRNAs that were linked by something that mimicked two amino acids with a peptide bond. This way, we would catch the ribosome right in the act of forming the bond. Though that project didn't work out, she abandoned her plans to return to the US after a year and then abandoned going to medical school, eventually spending ten years at the LMB before returning to the US as a faculty member at Caltech.

Luckily for us, *Science* was interested in publishing our release factor story, and we were able to rush through a reasonable paper on RF2 bound to the ribosome. The papers from Harry's lab on the topic, along with our own efforts, meant we were beginning to understand how ribosomes terminate at the stop codons at the end of a gene.

By then, though, several other labs had reproduced these crystals, which Maria had originally discovered. In addition to Harry's lab, Marat, too, had switched over to this new crystal form, and Tom's lab had also reproduced them. So we had blown our lead of almost two years with our magnesium blunder, and exploiting these crystals to get new insights into the ribosome had become a free-for-all.

However, if we wanted to address two of the biggest remaining questions about the ribosome, how tRNAs are delivered to the ribosome or tRNAs and mRNA move through it, these crystals were totally useless. These steps are catalysed by the GTPase factors EF-Tu and EF-G, which hydrolyse GTP as they make the ribosome machine chug along. In these crystals, a protein called L9 stuck out and bound to a neighbouring ribosome molecule in the crystal, in a way that would block the binding of any EF-Tu, EF-G, or, indeed, any other GTPase factor. So there was no way to use these crystals to solve complexes with the factors bound, and we were at yet another impasse.

A couple of years earlier, when we had first discovered the problem, I had been mulling it over and thought it would be a great idea to simply delete part of the gene for L9. We would thus eliminate the part of the L9 protein that sticks out and prevents the binding of the factors we were interested in. All excited, I went into the lab one morning and announced my brilliant idea. Frank laughed and with his usual sarcasm said that I was a little late. It turned out that Maria had already thought of this and had already ordered the necessary DNA fragments to genetically engineer exactly this deletion. She and Albert constructed a new strain of *Thermus* that lacked the protein.

Alas, when they tried to crystallize these mutant ribosomes in the same condition as ribosomes from the natural (or 'wild-type') *Thermus* strain, they couldn't see any crystals at all. We needed to find new conditions in which these mutant ribosomes would crystallize, preferably with one of the factors bound. Before he left, Frank had tried to crystallize the mutant L9-lacking ribosomes with EF-Tu, but nothing seemed to have worked.

Now a new postdoc, Yonggui Gao, wanted to give it a fresh shot. He had done his PhD in China and done a first postdoc in Japan with Isao Tanaka. He spent about a year trying to crystallize the mutant ribosomes with EF-G before he got some crystals in a completely new condition. When he collected a low-resolution data set, we could see that they were indeed in a new form, and moreover, they contained EF-G. At this point, Ann Kelley told us that she had looked at some of the crystallization trials that Frank Murphy had set up with the mutant ribosomes and EF-Tu just before he had left the lab. There were small crystals in almost the same condition as Gao's. We realized that in one stroke, we could have two key steps of the ribosome in the elongation cycle.

Solving both factors would have been far too much work for one person so I told Gao to focus on EF-G and asked Martin Schmeing if he could work with Rebecca and Ann on EF-Tu. Martin was an experienced crystallographer who had done some beautiful work as a grad student in Tom Steitz's lab on trying to understand how peptide bonds were formed in the ribosome. He was a tall, good-looking man with striking blue eyes, and was an odd mixture of a complete sports jock on the one hand and a hopelessly sensitive and moody romantic on the other. I had met him at meetings before and persuaded him to come and work on ribosome structure using electron microscopy in my lab, so he wouldn't be in direct competition with his old mentor Tom.

He had started work in my lab with a fellow Canadian, Lori Passmore, who had come from David Barford's lab in London, where she had on her own initiative learned electron microscopy by collaborating with Wah Chiu in Houston. She had originally written to Richard Henderson about a postdoc, but I felt I could use her skills to get our lab into electron microscopy and asked her if she would consider my lab instead. Ditlev and I had dabbled a bit in the technique, but it was really a full-time job that required someone who knew what she was doing. Luckily, Lori agreed.

The two of them were among the brightest postdocs I have had and worked for a while on how ribosomes start translation in eukaryotes. But after some initial success, Lori had gone on to start her own group at the LMB, leaving Martin to work alone on what was turning out to be an intractable problem. By the time I approached him to see if he would be interested in switching to the crystal structure of EF-Tu bound to the ribosome, he had already become hugely depressed at his lack of progress with electron microscopy.

It took him only about a day to jump at the chance to do crystallography again, especially on a project that he had already been interested in ever since he was a grad student at Yale. He teamed up with Rebecca and Ann Kelley to get cracking on it, while Gao plugged away at EF-G. It had just turned 2009, and we were back on track again.

CHAPTER 18

The Phone Call in October

THE YEAR 2009 BEGAN WELL. The new crystal forms of the ribosome with EF-Tu and EF-G were steadily improving. Pretty soon, we had workable maps of both structures, and as they were gradually being built, we had two new snapshots of the ribosome in action. To a great extent the broad outlines of what to expect were already known from electron microscopy, both from Joachim Frank's lab and those of some his protégés, like Christian Spahn in Berlin. But as the maps improved, we could see atomic details of what was happening, including all the subtle movements in both the factors and the ribosome, as the machine went through its cycle. It was very exciting.

The politics of the ribosome maintained a steady drumbeat even in the midst of exciting science. By now I had received lots of recognition for our ribosome work over the years. I had been elected to both the Royal Society in Britain and the National Academy of Sciences in the US. In 2007, I was awarded the Louis-Jeantet Prize for Medicine. This was a prestigious prize, but it was restricted to active scientists working in Europe, which ruled out the other major players in the ribosome. It also emphasized the 'active' by earmarking the bulk of the prize money for research rather than personal use and generally ruling out people near the end of their careers.

However, international prizes for work on ribosomes always seemed to go to other people. Deep down, I felt that the scientific event that transformed the field more than anything else was the determination of the atomic structures of the ribosomal subunits and the functional studies that followed as a result, to which we had made a major contribution. But it was clear that most juries did not see it that way. I had already accepted that I would probably not get a major international prize for the ribosome, but I confess I felt some trepidation each October. Every time I learned the Nobel Prize was for something other than the ribosome, I felt relieved because it was a postponement of the inevitable disappointment. What was invidious was that even if you didn't really care about the prize itself, by plucking out no more than three from a cohort of people who had all done important science, they made the rest feel like also-rans.

As the years went by, various people who had taken part in the 2004 Tällberg meeting got the nod from the Nobel committee. One of the first was Roger Kornberg, who got the prize all by himself for his work on eukaryotic RNA polymerase, the large enzyme that copies DNA into RNA in higher organisms. Nobody questioned that Roger deserved it, but it was odd that they had given it to him alone. If they felt that eukaryotic transcription was particularly important, they could have also awarded the prize to Bob Roeder, who had discovered the three distinct RNA polymerases in higher organisms. Or if they felt that the structure and mechanism of an RNA polymerase were important, they could have also given the prize to Tom Steitz, who had solved the first RNA polymerase structure (and, even earlier, a DNA polymerase), as well as to Seth Darst, a former postdoc of Roger's who had solved the RNA polymerase from bacteria, which had a conserved core with those from higher organisms. My guess is that the Nobel committee couldn't agree on the others, so they simply gave it to Roger. Notably, a year earlier the Lasker Award in the same area had been given to Bob

Roeder rather than Roger – which shows just how subjective these prizes can be.

A couple of years later, Roger Tsien, another speaker at the Tällberg meeting, shared the prize for his work that had transformed the way we can see structures in cells by labelling them with fluorescent proteins. It seemed to me and many other scientists that there might never be a Nobel Prize for the ribosome because the problem of choosing no more than three people out of all the contributors appeared insurmountable.

That was exactly the problem that a distinguished scientist on the Lasker jury wanted to solve when I ran into him on a visit to the US early in 2009. Apparently, they had spent year after year discussing the ribosome at length without agreeing on who should get it because of their self-imposed limitation of at most three people. He wanted my opinion of the various contributions to help him break the impasse. I protested that I was hardly likely to be objective, but he said he would take that into account.

I told him I couldn't talk about my own contributions, but we had a frank chat about the others. There was no question what Tom Steitz and Peter Moore had done, I said. However, after the initial Rosenstiel Award, Peter too was left out of the other prizes. This is largely because he was at the ribosome end of the collaboration with Tom, but by the time they started the project at Yale, Ada had already produced the well-diffracting crystals that they used as a starting point. So it was a crystallographic problem, which was largely Tom's domain. In fact, over the years, Peter had gracefully decided he didn't want to muddy the waters and had begun to support Tom for awards.

Harry Noller had spent his life relentlessly focusing on the important questions about the ribosome and, more than anyone else, had led the way to the thinking that the ribosome was fundamentally an RNA-based machine. His sequencing of ribosomal RNAs had led Carl Woese to discover a third branch of life, the archaea. But much of the biochemistry of the ribosome – such as which subunit read the code and which carried out the peptide bond formation, the discovery of the tRNA binding sites, and the role of factors – was all established before Harry began his work. A lot of Harry's biochemical work had involved trying to measure proximity between various components in the ribosome by biochemical methods, and these didn't really lead to a major insight into how the ribosome actually worked and were rendered obsolete by the crystal structures. The role of RNA in the ribosome was first suggested by Crick, and although Harry did many experiments that were strongly suggestive that ribosomal RNA was important, the fact that the ribosome was a ribozyme was conclusively shown by other people's high-resolution crystal structures. By then, catalysis by RNA had already been discovered in other contexts by Cech and Altman. As far as the structures went, the original whole ribosome structure from his lab was at low resolution and derived from the atomic structures of the two subunits done in other labs.

When we discussed Ada, he felt that her time had come and gone. I replied that prizes should consider who made a pioneering contribution that helped to start a field, and she should not be counted out simply because she had stalled and events had passed her by. It was Ada who had the vision to realize what was needed to crack the problem of understanding the ribosome and had not only done the initial work that set the stage but also kept the project alive for over a decade before producing crystals of the 50S that diffracted to high resolution. It had been done originally with Wittmann, but he was now dead. I told my interlocutor that she should not get a prize alone either, since it was others who had brought in new approaches and seen it through, as well as carried out detailed functional studies on both subunits and whole ribosomes.

After our conversation, he asked me to send him a written summary of my views. Although I was no longer optimistic about myself, I couldn't resist including an annotated copy of my CV in my report – just in case. Neither he nor I realized it at the time, but by then it was pointless. The distinguished scientist had waited too long to act.

There was no time to fret too much about it because there were papers to write. In addition to the two papers we were preparing on EF-Tu and EF-G for *Science*, Martin and I were also writing a long review of the ribosome for *Nature* that was now overdue by almost a year, and we were furiously writing away.

No sooner were the manuscripts submitted than I had to go to a meeting in Cold Spring Harbor. That year's symposium was on evolution at the molecular level, to celebrate the 150th anniversary of the publication of Darwin's *Origin of Species* and the 200th anniversary of his birth. I was the token ribosome speaker at a session on the chemical origin of life, much of which was devoted to how life could have arisen from an RNA world. I was a bit surprised to be chosen to speak, since I was not an RNA aficionado like Harry and the RNA aspect of the ribosome was only incidental to our work. My talk was on how all the key sites of the ribosome were entirely made of RNA, including the ones that recognized mRNA, made peptide bonds, and bound tRNA. The rest of the session had bonafide RNA gurus like Tom Cech, Jerry Joyce, and Jack Szostak, who covered the gamut from how RNA might start replicating itself and how proteins might gradually have replaced RNA and evolved into today's enzymes, to how the first cells might have formed. A ringer in our session was Craig Venter, the maverick scientist who was famous for his genome sequencing work as well as trying to construct artificial genomes. He flew in just for his talk and left immediately afterwards – clearly, he lived in a different world from the rest of us.

I noticed Jim Watson sitting in the front row during my talk. Afterwards, Jerry Joyce said Watson had come in only for Venter's and my talks. 'You must be on somebody's short list,' he quipped. Not on Watson's, apparently. After the session, I met him in the lobby where coffee was being served. He quizzed me about who was doing what in the ribosome and specifically asked what Harry was up to. Then he paused, peered at me intently with his bulging eyes, and said my work was beautiful but I really shouldn't worry about Stockholm because not getting the prize was not the end of the world. It was so blunt and completely unsolicited in the way that Watson was famous for that I was amused rather than outraged by his assessment. Perhaps he'd forgotten that he'd already said exactly the same thing to me the first time we met on the plane ride nine years earlier.

Soon afterwards, in August, there was a small meeting in Cambridge to which Ada was invited. She asked if she could visit the LMB, so I arranged for her to give a talk, to the great amusement of some of my colleagues who were aware of the tension between us. I introduced her by saying that while we might argue about who had done what in the ribosome, there was no question about who had started the crystallographic work and recounted the story of my first encounter with her at Yale in what seemed an age ago. Ada had taken to dressing almost exclusively in black, so I had no doubt what she would be wearing. By then, I too had grown fond of an all-black attire, which felt like a suitable homage to Johnny Cash, who had been so helpful to Bil while he froze so many crystals.

So as a joke, I came dressed in black, and Martin Schmeing took a picture of the two of us in our matching outfits outside my office in front of a poster for the Stetten Lecture at NIH where Ada and I had both spoken in 2000. We ended the day by taking Ada out to dinner at my favourite South Indian restaurant in Cambridge. The tensions of the past seemed behind us, and we all enjoyed chatting and joking about science.

By this point, it was the end of September. Our papers in *Science* and *Nature* had been revised, and Martin was working on trying to submit an aesthetically pleasing possibility for the cover of *Science*. We were all in a great mood. I had to take another short break, this time to go to Vienna, where I was on the advisory board for the Institute of Molecular Pathology or IMP. I had been doing it for a couple of years, and it was always a pleasant exercise to listen to scientists doing outstanding work talk about their progress. I'm not sure my advice was worth much, but it was also an opportunity to get to know prominent scientists from all over the world who were part of the committee.

Figure 18.1 Ada Yonath and the author in all-black outfits

One of them was Eric Kandel, the famous neuroscientist from Columbia University who had done a lot of work on the basis of memory. As was often the case on these occasions, someone brought up the 'ribosome prize.' Kandel, who was part of the conversation, mentioned that he definitely thought it was worthy of a Nobel. He revealed that he was on the Lasker jury and that there was always a major discussion around it. He then told me I was always central to the conversation but hastily added – in case I got my hopes up – that I shouldn't count on it! I laughed and assured him that I was not holding my breath. At this point, Barry Dickson, the director of the IMP, asked me what I thought would happen with the ribosome Nobel. I joked that it was not worth worrying about yet because some of us would have to die first.

The Nobel Prizes were announced the week after I returned from Vienna. They always follow a set order in early October. On Monday, they announced the prize for physiology and medicine, which went for the discovery of an RNA-based enzyme, telomerase, which extends the ends of our DNA and prevents them from shortening too much during our lifetime. One of the recipients was Jack Szostak, with whom I had shared a session a few months earlier at Cold Spring Harbor. I knew he'd be deluged with messages, so I congratulated Jon Lorsch and Rachel Green, who had both been students of Jack's and said their odds had gone up because statistically the surest way to get a Nobel Prize was to have worked for someone who had won it. Since Rachel had been Harry's postdoc, I told them it was possible that eventually Rachel might double her chances.

The prize for chemistry was to be announced on Wednesday. The chemistry prize often alternates between the hard-core chemists and the more biological chemists. This was often a cause for contention among the hard-core chemists, who understandably complained that the chemistry prize was frequently awarded to people who barely understood any chemistry. Since they had given it to a 'biological' area the year before, I figured the ribosome would not be a candidate and I would be reprieved for another year. So by Wednesday morning, I had completely forgotten about it. Halfway to work, I got a flat tyre on my bicycle and had to walk the rest of the way.

I came into work late in a grumpy mood when the phone rang. I answered curtly, only to hear a woman at the other end saying that it was an important phone call from the Swedish Academy of Sciences and could I please hold the line. I immediately suspected that this was an elaborate prank orchestrated by one of my friends like Chris Hill or Rick Wobbe, who enjoyed practical jokes. Chris had once written a letter to Guy Dodson, who had chaired the official interview committee for my LMB job, saying it was 'good that you were able to do something to help Venki out since he was having trouble here in Utah but might do OK in the less competitive environment of Britain,' which had caused Guy to call me in Utah in a panic to find out if I had pulled a fast one over him and the LMB.

Finally, a Swedish scientist who introduced himself as Gunnar Öquist came on the line and said I had been awarded the Nobel Prize in Chemistry along with Tom Steitz and Ada Yonath for our work on the structure and function of the ribosome. When he finished his spiel, there was a slight pause. On the one hand, that was the only combination of recipients that could possibly include me and meant they had recognized that the atomic structures were what had changed the field. But I still found it hard to believe – especially given my previous altercation with Måns Ehrenberg at the Tällberg meeting and his subsequent appointment to the Nobel committee. So I told him that I didn't believe him, even though he had a very good Swedish accent! At this point I could hear laughter and realized I was probably on a speakerphone at the other end. If it was true, then Måns Ehrenberg ought to be there, and I asked if I could speak to him. There was more laughter, and then Måns came on the line. He congratulated me and said I deserved it, but it was the last time he was going to say it! Then, perhaps sensing my ambiguity about prizes, he pointedly asked, 'You're going to come and accept it, aren't you?' Suddenly I knew that it was true after all. People often ask how I felt when I first found out, but the full extent of it sank in only gradually. It certainly did not bring the immediate burst of excitement that made Brian and me do high fives at the Brookhaven synchrotron when we saw peaks from the tungsten clusters, or when we saw our anomalous peaks in Argonne and knew the structure had been cracked.

Occasionally when people would ask me about the prize, I would jokingly blow it off by saying who would want to go to Sweden in cold, dark December to eat bad vegetarian food. Sometimes, I had fantasized about refusing it. But the reality is that no matter how people may feel about prizes in the abstract, it is very hard to actually turn them down, especially something as grand as the Nobel Prize. It is enormously satisfying to know that your fellow scientists think so highly of you. It is also a great tribute to the students, postdocs, and staff who had risked their own careers on the project and without whom nothing would have been accomplished. And of course, the cash is always welcome. Even Richard Feynman, who disdained awards, had accepted it.

I realized then that Måns was a man of real integrity. He had obviously put aside his differences with me about one aspect of our work and considered the bigger picture. At that level of discussion, even a slight lack of enthusiasm can sink a candidate. Had he been even slightly vindictive, he could easily have sidelined me from consideration and nobody would have known. Perhaps it is because of the integrity of people like him that the Nobel, despite all the controversies it frequently engenders, is still so highly respected.

Afterwards, Anders Liljas and Gunnar von Heijne also came on the line to congratulate me. Finally, I was told that I could tell my wife, but I should not tell anyone else until the official announcement. In the meantime, they said, I should enjoy my last thirty minutes of peace.

Unknown to me, Martin and Rebecca, whose desks were right outside my office, were listening in. They shared none of my scepticism – in fact, a year earlier, Martin had bet me a dinner that I would get the prize. By the time I had hung up, they were jumping up and down in excitement. Martin popped open a bottle of champagne that he had been saving to celebrate the *Science* papers when they came out.

I tried to call Vera, but there was no answer. She had gone on a walk with my stepdaughter, Tanya, who was visiting from Oregon, and since she didn't use a mobile, there was no way to get hold of her. When she returned, a good friend, Peter Rosenthal, called her. He was a Harvard student with Don Wiley before he came to do a postdoc at the LMB with Richard Henderson and now works in London. He had a deep voice and the measured professorial manner befitting a Harvard man. He told Vera he thought it wouldn't be possible to reach me at work, which is why he was calling her at home. Vera was puzzled. She never had any trouble reaching me, she replied. Peter paused and said, 'Is it possible that you don't know?' 'Don't know what?' asked Vera. And that was how she found out. Her response when we met later that evening was, 'I thought you had to be really smart to win one of those!' To quote Maryon Pearson, the wife of the former Canadian prime minister, 'Behind every successful man, there stands a surprised woman.'

For a relatively small institution, the LMB has received so many Nobel Prizes that a reporter referred to it the next day as a Nobel factory. Aaron Klug pointed out that a farm or garden was more appropriate – we plant seeds and nurture people and the Nobels are if anything an occasional side effect of good science. But nevertheless, a tradition had grown up over the years about celebrating the Nobel. Mike Fuller, who as a teenager had been one of the first people to join the LMB's staff, organized the customary champagne celebration in the canteen, as he had for all the previous Nobel Prizes awarded to scientists here over the decades. By the end of the day, a steady stream of people had made it to the top floor where the canteen was. There were lots of photographers, and I regretted that on that day of all days, I had forgotten to shave and looked quite dishevelled. One journalist wanted to take a picture of me with my lab members and stuck a glass of champagne in my hand and asked me to hold it up, which amused Daniela Rhodes greatly because she knew I was almost a teetotaller. I was soaking it all in, feeling a mixture of happiness and relief that it had all worked out – the worries and stresses at the start now seemed to belong to another world. Although the science had happened over many years, it was also a moment for the LMB, especially Richard Henderson, to feel vindicated in the gamble they had taken on me. After the celebration, Vera and I walked my bicycle home in the rain.

CHAPTER 19

A Week in Stockholm

THE TWO PAPERS IN *SCIENCE* had come out, and we had made the cover, with a composite of the ribosome in both states. At the same time, our review had come out in *Nature*, with the cover announcing that it was by this year's Nobel Prize winner. Coming on top of the prize, it seemed almost too much. My cup was not just full, it was overflowing. I felt nobody had the right to so much success all at once.

The Nobel people were right about my thirty minutes of peace. Once the prize was announced, the phone didn't stop ringing for two days and had to be diverted to the central switchboard of the LMB. I was particularly thrilled to talk to the *New York Times* and NPR, which I read and listened to religiously even after moving to England. Curiously, although our friends from Grantchester saw the news on French television, I did not make the evening news on British television or even the print versions of many newspapers.

Instead, there was a deluge of calls from journalists in India. I had left India when I was nineteen and had largely been ignored there except by people in my field. Suddenly, I became the subject of an entire nation's celebration. There was the usual agonizing among some pundits there about whether it would have been possible to win the prize by working in India, something that had not happened since colonial times when C. V. Raman won the Nobel Prize in Physics. I was pleased but not entirely surprised to get congratulatory letters from President Obama and Prime Minister Gordon Brown because I was an American citizen living in Britain. But I was surprised to get letters from the Indian president and prime minister as well, since by then I had not lived in India for almost forty years and had not been an Indian citizen for most of that time.

Scientists are not used to public attention, and when a flood of emails from random strangers in India continued unabated for days, I was annoyed. When a journalist asked me if it was true that I had been offered the directorship of an institute in India, I snapped that I hadn't and wouldn't accept it anyway, and then complained about random strangers from India clogging up my in-box and affecting my ability to carry on my work. It was a real-life lesson in media training because the next day my complaint was on the front pages of every major newspaper in India. The adulation turned to fury, and now I got angry emails from strangers denouncing me for forgetting my roots and being arrogant. A contrite clarification partly mollified some but further infuriated others because I had said nationality was an accident of birth. Some Hindu nationalists had already been annoyed with me because they had learned from news reports that in the aftermath of the Gujarat riots in 2002, I had supported a scholarship to help poor Muslim girls, partly as a gesture from someone who was of Hindu ancestry and partly because the education of girls lifts society as a whole everywhere. Now they had further reason to think I was a traitor to their cause.

On the other hand, one of the nicer consequences of the prize was to hear from lots of colleagues and friends, including some with whom I had lost touch over the years. One of the first to write was Barry Dickson from Vienna, who reminded me of our conversation only a few days earlier and said he was happy that nobody had to die for it to happen. Peter wrote to congratulate me and said that he had told Watson a few years earlier that exactly this combination should get the prize. He went on to say that he was very pleased that someone who had once worked with him would go on to have the career I had. I thought it was particularly generous of him to say so, especially under the circumstances, and it reminded me again of his fundamental courtesy and decency.

Pretty soon, it was time to head off for Stockhom. The Nobel Foundation and the Swedish Academy of Sciences put on a show that lasts almost a week. The whole thing is a great PR effort on their part to protect the Nobel brand – to make you feel so special and the occasion so memorable that there is no confusion about which prize is the 'real' thing, especially now that there are other prizes that offer a lot more money.

The week starts off with the Nobel Foundation assigning you a personal assistant and a chauffeur for your entire stay. My assistant was a young man from the foreign ministry with the very Swedish name of Patric Nilsson. He said he would meet Vera and me at the airport and take us to the Grand Hotel. I told him that I had flown into Arlanda Airport and taken the train to central Stockholm many times and he didn't need to meet me at the airport on a Saturday evening just to take us to the hotel. He politely insisted, and I relented.

Clearly, I hadn't understood. When Vera and I disembarked, we were greeted right at the door of the plane by none other than Gunnar Öquist, he of the 'very good Swedish accent.' Next to him was a young Indian man, who introduced himself as Patric Nilsson! He had been adopted from India as a baby and, of course, was stereotypically Swedish in every aspect except his skin colour. They apparently thought it would be amusing to pair us up. Then came another surprise. In the jet bridge connecting the plane to the building, there is a door right near the plane that I had ignored my entire life. Vera and I were whisked through that door down some stairs to a waiting car and then to a VIP lounge, where all the immigration formalities and our baggage were taken care of while we sat around and chatted with Gunnar. It gave me a taste of how the

Figure 19.1 The author's lab members over the years at a rooftop restaurant in Stockholm

very rich must live and why we don't see them in immigration lines; and how, as F. Scott Fitzgerald said, 'they are different from you and me.' The next morning, I realized I had forgotten to pack a tie, and Patric brought me a selection of his, from which I chose the least flamboyant – a Scottish tartan.

The Nobel people said I was allowed to bring about a dozen guests in addition to Vera, so I invited our children, Tanya and Raman, my daughter-in-law Melissa, and my sister Lalita and brother-in-law Mark Troll. I also invited my good friends Bruce and Karen Brunschwig – after all, without Bruce's osmium hexammine, I wouldn't be in Stockholm. The remaining slots went to the students and postdocs who had risked their careers to work on the 30S subunit. A final slot went to Richard Henderson as gratitude for hiring me at the LMB and his years of support. This may have been my 'official' delegation, but Bil, ever the social organizer, was not one to be inhibited by the official limitation on the number of guests. He took charge and organized a parallel celebration in Stockholm consisting of nearly everyone who had worked in my lab. He even got suggestions from the Nobel people for where they should stay and where we should get together for parties. One of those occasions was lunch at a terrific vegetarian restaurant in Stockholm with all of them and Anders Liljas, so I can never again complain about bad vegetarian food in Sweden.

On another evening, we had dinner on the top floor of a building with a great view of Stockholm, and members of my lab roasted me with funny incidents they remembered about me during their time in the lab. They brought up my suggestion of the guillotine, which destroyed two hundred crystals, my locking us out of two cars simultaneously during a synchrotron trip, my once inadvertently deleting some data we had just collected at a synchrotron, and many other embarrassing episodes.

The official Nobel programme filled the rest of the time with lots of dinners, receptions, and interviews. To me, the most important of these were the Nobel Lectures, held in one of the largest auditoriums at Stockholm University. I had read many of the Nobel Lectures ever since I was an undergraduate, and they were often so beautifully historical, scholarly, and coherent that I was daunted at having to give one. Accordingly, I had prepared a careful talk on how the ribosome maintains the accuracy of reading the genetic code. I hadn't realized that the Nobel Lectures I had read were not always a transcript of their actual talks. It came as a complete surprise when Anders Liljas told me that my talk would be for a large and quite general audience of students and faculty from Stockholm University. So I stayed up until about 1 a.m. redoing my presentation and making it more generally accessible. The next morning, I was too exhausted to be nervous.

We were to speak in alphabetical order, and unusually for someone whose last name starts with an *R*, I was relieved to be speaking first. I began with a picture of all of the people in my lab who had contributed to the ribosome work – nearly all of whom were actually sitting in the audience. I pointed out that both Paul Zamecnik and Mahlon Hoagland, who had discovered tRNA, had died in just the previous few months and showed a clip of Jim Watson talking about the early days of ribosome work. Then, after describing our work on the structure of the ribosome and what it told us about decoding, I finished by showing a movie made by Martin and Rebecca of the ribosome during decoding. They had taken the

Figure 19.2 The author, Tom Steitz, and Ada Yonath at the Nobel Lectures. On the left is the moderator, Gunnar von Heijne.

metaphorical movie we were constructing by taking snapshots of the various states and assembled an actual movie by interpolating the movement of the ribosome and tRNA between those frames. Then, following something Martin had done for Tom when he was a student at Yale, he and Rebecca decided it would be good to set it to music. Unlike the classical music Martin had used for Tom, they decided to use rock songs with lyrics that were related to what was happening. So for example, when the tRNA was sampling the codon and could stay bound or dissociate, Mick Jones from The Clash sang 'Should I Stay or Should I Go.' Or when the ribosome's bases interacted with the minor groove of the base pairs between the tRNA and codon, Madonna started singing 'Into the Groove.' When the tRNA finally entered the peptidyl transferase centre to deliver its amino acid, the movie concluded with a triumphant 'We are the champions of the world' by Queen. I had never heard of most of the songs or even the groups, but the movie was a big hit.

Over the next few years, people would come up to me after my talk and instead of asking about my work they would ask if the movie was available to download.

I could now relax and hear Tom and Ada speak. Their talks were so familiar to me now from the years of the ribosome road show that Tom and I used to joke that we could have given each other's talks, but there was sometimes something new and unexpected. Tom gave a polished talk focusing mainly on how they had solved the 50S subunit, the peptidyl transferase reaction, and the various antibiotics that bound to the 50S subunit. He too showed a movie made by Martin on how the peptide bond is formed. At one point, he showed a picture of Peter, catching a big fish, saying the ribosome was certainly a big fish. Peter was in the audience as Tom's guest and collaborator, and I again felt a bit guilty that he, my mentor who had introduced me to the ribosome, was not sharing in the credit for the work.

Finally, Ada spoke. Her talk had the intriguing title 'Polar Bears, Antibiotics and the Evolving Ribosome.' I had no idea why polar bears were part of the title but soon found out. She said that when she was hospitalized after an accident, she read in a magazine that hibernating polar bears organize their ribosomes in crystalline arrays, and that was what originally gave her the idea to crystallize ribosomes. After all, she said, if polar bears could do it, why couldn't we do it in the lab? She then went on to discuss the early days of ribosome crystallography, showing a picture of Håkon Hope at the beamline when they tried to flash-cool the crystals, following it up with her 30S and 50S structures and her antibiotics work. It became apparent that she had already settled into her role as a Nobel laureate because she ended her talk with some speculations on the first amino acid and the origin of life.

When her talk finished, I was a little puzzled because I had never heard about polar bear ribosomes in all the years I had been working in the field, nor had anyone else I knew. There was also no mention of it in Elizabeth Pennisi's article on the ribosome race in *Science* in 1999, for which she had talked with Ada herself about how the crystallization effort had started in Berlin. Anyway, by the time I joined Peter's lab in 1978, at about the same time Ada began her work with Wittmann in Berlin, it was already well known that ribosomes could form ordered arrays. I had even heard Nigel Unwin give a talk about his work on two-dimensional crystals of ribosomes from lizards in Washington, D.C., in 1977 when I was on my way to my interview in Peter's lab at Yale.

I didn't think much of it until a couple of weeks after I returned, I got an email from Don Engelman, who informed me that polar bears do not hibernate. Apparently, only pregnant females enter a den to have their cubs, but even then they are not in a state of true hibernation. The polar bear story popped up again at a meeting in Erice the following summer. When Ada brought it up again, Tom pointed out that polar bears do not hibernate. This led to an altercation between the two of them, which ended with Ada saying that it might well have been some other kind of bear. Some in the audience seemed to like the polar bear story and were not happy that Tom had questioned it. Intrigued, I searched the literature for *any* kind of bear that was the source of ordered ribosome arrays and found nothing. Despite Tom's questioning its premise, she has often repeated the story over the years, and audiences seem to love it. In any case, the provenance of the polar bear story is still a mystery. I cannot help picturing some intrepid biochemist crawling into a den in the snow to extract tissue from a polar bear resting next to her cubs.

One of the events of the week was a panel discussion for a BBC television show called *Nobel Minds*. The host briefly asked us about our work but then quizzed us about Obama's peace prize, climate change, and all sorts of issues. This experience gave me a glimpse of my future life in which Nobel laureates are treated as though they are sages and asked to pontificate about topics well outside their area of expertise.

The prize ceremony itself is always on December 10, the anniversary of Nobel's death. We were all dressed in tails or gowns and made to wait until we walked in procession onstage. I waited anxiously backstage, knowing my daughter-in-law Melissa had not yet arrived from the US. When we were finally steered into the auditorium, I was relieved to see that she had made it with just minutes to spare before the doors had closed, and we flashed each other a big smile. The royal family arrived, and the speeches – in Swedish – began, interspersed with music from the Stockholm Philharmonic. We got our certificates and medals from the king, bowed, and returned to our seats. The king, who has been doing this year after year, understandably looked bored.

The climax of the evening was the Nobel banquet, which oddly is televised in Sweden and webcast live. I wouldn't have thought spending the evening watching a group of strangers eat would make for much drama, but there were speeches and entertainment. Each of us was assigned a place of honour at the long table in the centre of a grand hall. We were to make an entrance with a companion down a grand staircase and take our place, while observed by the rest of the guests who were already in place. Vera, who was seated a few places to my right, was accompanied by Germany's vice chancellor. Ada was leading the procession, accompanied by the king, to be followed by the queen and the head of the Nobel Foundation. I was next, and my companion was Victoria, the crown princess of Sweden. She was clearly a favourite of the press, so as we proceeded to the hall, there were a blinding series of flashes from cameras and I could barely see my way down the stairs. Eventually, we sat down, with Ada on my left and the princess on my right. On the other side of the princess was Tom. It turned out that she had spent some time at Yale, and Tom and I asked her about that. But conversation with the princess was not as easy as talking to Ada, with whom I had a lot more in common and who has always had an ironic wit and sense of humour.

There was some great live entertainment with Renaissance music and period costumes. Towards the end of the dinner, each category of prize winners had one of their group go up and give a short speech. Following tradition, Tom and I asked Ada, the most senior of us, to give the speech on our behalf. Along with describing the challenge of the ribosome at the start, she mentioned how in Israel her curly hair meant she had a head full of ribosomes and how the polar bears that had inspired her were now threatened by climate change. She ended by profusely thanking her chauffeur for the week but did not mention Tom or me by name.

After the dinner, as we got up, I clumsily stepped onto the long train of the princess's elaborate gown, but just as we were about to start walking, she noticed and yanked it free from under my feet with a single expert movement of her hand. We then said goodbye to the royals with whom each group of laureates gathered in turn for a photo op, but the party went on, with dancing and just hanging around and chatting. The evening continued at an after-banquet event at Stockholm University until late in the night. The next morning, several newspapers had front-page photos of Tom and me flanking the princess at the dinner. It was very gratifying to see that the press had grasped the fundamental importance of ribosomes.

Vera and my other guests left a day later. I skipped a last banquet held by Stockholm University because I had a much better alternative: Måns Ehrenberg had invited me and a number of ribosome people to a dinner at his home in Uppsala after my talk there. After all the formalities of the week, I could finally unwind and relax with old friends like Anders Liljas and Maria Selmer, my former postdoc who was now a faculty member there. Then, after a talk the next day in Lund in the south of Sweden, it was suddenly all over.

CHAPTER 20

Science Marches On

THE SUDDEN DECOMPRESSION AFTER BEING in the limelight in Stockholm, combined with the dark winter in Cambridge, was anticlimactic. It didn't help that I remembered that many years earlier, Anders Liljas had told me the Nobel Prize was the kiss of death because the subsequent distractions kill off people's research – in other words, they destroy the very thing that made them known in the first place. But in Stockholm, Gunnar von Heijne gave me a useful bit of advice. He said it was entirely up to me how I should live my life, but if I wanted to keep doing science, I could do worse than emulate Rod MacKinnon, a structural biologist at Rockefeller University whom I knew well. Rod is an intensely focused scientist who has not let the Nobel Prize slow him down and continues to make important advances by refusing to be distracted by the inevitable invitations that follow the prize. I was determined to prove Anders wrong, so Rod became my model.

Our immediate task was to keep plugging away at making more frames of the ribosome 'movie.' But the remaining states were getting harder to trap and even harder to get good crystals of. Although we had the occasional success, I could see that it would be increasingly difficult to persuade talented people to come and work on the missing frames of a movie that was largely taking shape, especially when there was no guarantee they would succeed even after years of work.

What I had not anticipated was that for the first time in about fifty years since Max Perutz and John Kendrew got the first glimpse of a protein, we would have a new method to visualize large biological molecules in atomic detail without having to crystallize them. When Joachim Frank had shown his maps of the ribosome at that fateful meeting in Victoria in 1995, we had all been impressed, but none of us had imagined that electron microscopy would ever give us the level of detail needed to deduce the atomic structure of the ribosome. We had dismissed it as blobology – good for a first look but not much more than that. Sure enough, it was crystallography that cracked the ribosome structure a few years later.

However, in the same year as the Victoria meeting, Richard Henderson – the man who gave me my job at the LMB – published a quite remarkable result. Electrons had the right wavelength, and physicists and metallurgists had obtained atomic structures with the electron microscope for decades. But you couldn't get high-resolution structures of biological molecules from electron microscopy because they don't have enough contrast, and if you hit them with enough electrons to get a signal, you end up damaging them. Richard calculated that despite this, if you could improve both the microscope and the detector, it should be possible to get an atomic structure using the method – without any crystals at all.

It took a long time to get there from 1995, but over the years, first the microscopes improved and then several groups developed new detectors that were faster and more sensitive than the film that had been traditionally used. One of the new detectors was developed by Richard and his collaborators and was installed in a microscope at the LMB in 2011. Several people, including my colleague Sjors Scheres, developed software to take advantage of the data from these detectors.

As a result, it is now possible to get maps that are just as good as our crystallographic maps. We were able to use this for all kinds of projects that had stalled for years. And the amazing thing is that you don't need crystals, which cuts out years of uncertain work. Moreover, you only need a tiny amount of material, and just as importantly, the sample doesn't have to be absolutely pure. Suddenly, it became straightforward to do ribosome structures, even very complex ones, so much so that the field has become a free-for-all. For a long time it had been impossible to do the structure of the mitochondrial ribosome by crystallography, but now on two separate occasions, Nenad Ban's group and my group published structures only a day apart.

It wasn't just the ribosome field either. All sorts of biological complexes that previously seemed impossible to solve because they were short lived or hard to get in large amounts or existed in multiple conformations could now be solved to near-atomic resolution without any crystals. In addition, it is now becoming possible to directly see molecules as they exist inside cells, revealing their organization in unprecedented detail. A new revolution in visualizing biological molecules is under way, with exciting new structures being reported on an almost weekly basis.

Looking back at the long struggle to get the first structure of a ribosome by crystallography, it is ironic that today it would only take a week or two from start to finish. The field is now flooded with structures of new types of ribosomes in all sorts of different states, and I can imagine journal editors groaning on receiving a YARS manuscript – yet another ribosome structure.

When the initial crystal structures were unveiled in Copenhagen in 1999, a lot of people were worried that their field had ended. They were only partly right. A lot of biochemists had been tediously trying to figure out which parts of the ribosome were close to which other parts, in an effort to piece together its structure indirectly. Once the atomic structures came out, these people had to find something else to do, as did scientists who were trying to solve the structures of isolated pieces of the ribosome. But those who were using biochemical methods to study how the ribosome worked found their work transformed because the structures meant the ribosome was no longer a black box. Geneticists and biochemists could modify the ribosome at will and interpret the functional changes quite precisely because they knew exactly where the changes were in the structure. It was gratifying that we had played a part in taking ribosome research to a different level where much more sophisticated questions could be asked.

Structures are static snapshots of a particular state of the molecule. So the so-called movie of the ribosome in action that we were trying to make was really a collection of stills of the molecule in different states. The stills can only suggest how the ribosome gets from one state to another and say nothing about how fast the transitions are, or whether there are states in between the stills that we just cannot capture.

Applying single-molecule physics to ribosomes has become an exciting new way of learning about these transitions. There were a couple of ways to do this. One of them was to attach fluorescent molecules to different parts of the ribosome or tRNAs. By measuring the resulting fluorescence in a technique called fluorescence resonance energy transfer or FRET, you could measure if the fluorescent molecules had moved relative to one another. With the structures in hand, the method could be used to attach the fluorescent molecules to precise points on the ribosome, and you could tell which parts of the ribosome were moving at which stage and how often and how fast.

The application of the method to single molecules of ribosomes was pioneered by Jody Puglisi, who was working with his colleague Steve Chu at Stanford. I already knew Jody from meetings. He was a strikingly good-looking man who reminded me of the Italian movie stars of the twenties and thirties. He had a sardonic sense of humour and would sit in the back of the lecture hall looking like he wasn't paying much attention to the talk, but would then raise his hand afterwards and ask the one killer question that exposed the weakest part of the argument. He had been working on pieces of ribosomal RNA and small proteins that bind to the ribosome, but once the structures of the ribosome came out, he very quickly realized that he needed to change directions.

By then, he had already started working with Steve Chu on applying single-molecule physics to the ribosome. I found out about their work when I went to Stanford to give a talk right after we had solved the 30S structure in 2000. Visiting an American university to give a talk has the character of a job interview. On your arrival, they give you a schedule with a list of faculty members to meet before you give your talk at the end of the day. It is interesting to talk to people working in different areas and learn about their work, but when you have travelled a long distance and are jet lagged, it can be a gruelling day. I didn't know Steve Chu or his interest in biology and was surprised to find the famous physicist on my schedule. I quickly looked him up and saw he had done some work on RNA folding. When Scott Blanchard – now a leading practitioner of the method but then a grad student of Jody's – escorted me to his office, Steve had clearly forgotten about the appointment. He greeted me with a nonplussed look, sat me down in his office, and asked me what I would like to talk about. I thought it was a bit odd, but said since I was interested in RNA, perhaps he could tell me about his work on RNA folding. He paused, looked at me, and said, 'Now let me get this straight. You're here to interview for a postdoc with Dan Herschlag, right?' His fellow faculty member Dan was a physical chemist working on RNA. I didn't know whether to be flattered that he thought I was young enough to be a postdoc (or perhaps he thought I was just slow), or to be insulted that he not only hadn't heard of our 30S structure but didn't know me at all! It was one of several occasions when I was brought down to earth just as I was feeling very pleased with myself.

Anyway, Jody and Steve, working with their students like Scott Blanchard and later Ruben Gonzalez, pioneered the use of the method for looking at ribosome function. This method is now helping us dissect in detail when factors arrive and leave, when the ribosome moves, and which steps are fast and which are slow.

A second physical method was even more amazing. Physicists had figured out how to trap single molecules in a field and exert forces on them. By doing this, they could actually do things like pull on the mRNA or the nascent chain and measure the force exerted by the ribosome when it translocates from one codon to the next. One of the leaders in this area is Carlos Bustamante at Berkeley, who teamed up with his colleague Nacho Tinoco (who sadly died recently) and Harry Noller, making a formidable combination with their complementary expertise.

So the structures were being used with a combination of old and new methods to understand how the ribosome functions as a molecular machine. But there are lots of questions beyond that. Sometimes cells need to make a lot of a particular protein. Sometimes they need to shut down production. What were ribosomes actually doing in the cell at any given time and how did the cell control their activity?

A new method to tackle this question had its beginnings a long time ago, when Joan Steitz showed that if you chewed away all the mRNA with a ribonuclease – an enzyme that degrades RNA – there was a piece that was protected by the ribosome that you couldn't chew away. This protected piece of mRNA is clearly where the ribosome sits on it. When Joan did this in the early 1970s, there was not much else you could do with it. But almost three decades later, Joan's finding was exploited in a revolutionary new way. Jonathan Weissman was a second-generation scientist in two ways – his parents were on the faculty at Yale, but later in life, his mother, Myrna, who later became a professor at Columbia, married Marshall Nirenberg, who had helped crack the genetic code. To cap it all, Jonathan told me he once did a project in Peter Moore's lab when he was a student. He was clearly steeped in molecular biology from when he was quite young.

Jonathan figured that with new sequencing techniques you could break open a cell, chew up all the RNA, and then take the pieces of mRNA protected by the different ribosomes in the cell, amplify them, and sequence them. You would then get a snapshot of what ribosomes were doing on every region of every mRNA at a particular instant. The technique, called ribosome profiling, led to all kinds of unexpected findings. You could see where ribosomes slowed down on a piece of mRNA, where they piled up, and where there were fewer ribosomes than expected. You could also see which mRNAs were being translated more and which ones less at different points in the life cycle of a cell. Suddenly, it became possible to ask detailed questions about what proteins the cell is making at a particular time and even in what amounts. The method has had a huge impact on understanding how ribosomes are used in the cell and where they encounter problems.

Beyond this, there are questions of how the cell regulates ribosomes and how viruses hijack ribosomes to translate their own genes. Cells also have sophisticated methods to stop the process when it goes wrong – there are lots of quality-control processes acting in the cell, which involve the ribosome in one way or another. Today, control of translation is involved in all kinds of processes from cancer to memory. There is now evidence that some of this control is brought about by specialized ribosomes, which would be ironic since they were crystallized in the first place on the assumption that they were all identical. Finally, scientists are using new and powerful tools to understand how the ribosome is assembled from its component parts in the cell and how this assembly is regulated.

For a while, the ribosome was thought of as an exceptional relic that had survived the transition from an earlier RNA world into our modern protein-dominated world. To everyone's surprise, people discovered in just the last two decades that there are lots of RNA molecules around in the cell whose very existence had not even been suspected previously. Some of them are very small RNA molecules called microRNAs, which control how genes are turned on and off. Sometimes they act on mRNAs, preventing the ribosome from initiating properly on them. At other times they can cause the mRNA to be degraded quickly. Sometimes they directly control the expression of particular genes by affecting how much mRNA can be made from the DNA. There are even longer RNA molecules that turn out not to code for proteins at all, and at least some of them are thought to control genes too. So the RNA world never really disappeared – it just evolved into a world where RNA acts in partnership with proteins to carry out the essential processes of life. Because of these new and completely unanticipated types and uses of RNA, the field of RNA biology has exploded.

As is nearly always the case in science, the structures of the ribosome simply moved the questions to the next level. When we have a clear goal in mind, we think we are struggling to reach a summit. But there is no summit. When we get there, we realize we have just climbed a foothill, and there is an endless series of mountains ahead still to be climbed.

EPILOGUE

DESPITE MY DETERMINATION TO STAY focused on the science, my life after the prize was different and not always for the better. Suddenly, I had been discovered. I was invited to appear on radio and TV and pontificate about anything remotely scientific and even about the future of the world. I was asked to give talks at meetings that had little to do with my own area of research. Lots of universities wanted to offer me honorary degrees (all of which I turned down except for Baroda, Utah, and Cambridge, where I had studied or worked), and learned societies and academies suddenly elected me as an honorary member.

Despite my comments to the Indian press that nationality was an accident of birth and they should leave me alone, the government of India decided to bestow upon me one of their highest civilian awards. I am not a big fan of ethnic and nationalistic pride (which seems to me the other side of the coin from racism and xenophobia). I am also not a fan of identity politics in general. When I grew up, some of my heroes were Indians, like the mathematician Srinivasa Ramanujan or the astrophysicist Subrahmanyan Chandrasekhar, but most of them were as varied as Richard Feynman, a Jewish boy from Queens, New York, and Marie Curie, a Polish woman who worked in France. It also did not matter that I had not met any of them personally; I could gain inspiration simply from reading about their lives and work. I am not sure that narcissistically taking a selfie with one of them would have provided any better inspiration. But I could see why I had inadvertently become a source of inspiration and hope for people in India: simply by the fact that I grew up there, studied at my local state university, and went on to do well internationally showed that you did not have to start off in elite places in the West to do well.

Somewhat later, in 2011, Britain decided to give me the honorary knighthood that they award foreigners, but when they found out I had also become a British citizen that January, decided to give me what they quaintly referred to as the 'substantive' version. Many of the great scientists at the LMB had turned down knighthoods, and I too felt ambivalent about accepting it. However, Vera said it was one thing for people born in Britain to turn it down, but it would be churlish of me, a guest in the country, to decline the honour. Also, even in 2011, I could see an anti-immigrant and xenophobic strain developing among some elements of the public. So I thought it would be good to show that immigrants come here and bring credit to the country, and was grateful to accept the honour.

Perhaps the most surprising of the honours was to be elected president of the Royal Society, one of the oldest scientific organizations in the world. They had promptly elected me as a fellow in 2003 as soon as I had fulfilled the residency requirements and long before I had become a British citizen. But becoming president of the Royal Society was very different. In effect, I would become a leading voice of British science. When I was contacted to see if I would be interested in the position, I was astonished, since I had moved to Britain fairly late in life and had spent all my life in the country within the confines of the LMB. I was certainly not a mover and shaker with extensive connections, nor had I headed large, important organizations; in fact, I had never even chaired a major committee before. Moreover, I'd had almost nothing to do with the Society since my election as a fellow. I thought I was an odd choice and certainly very different from some of my immediate predecessors.

I didn't know what to think. It was hard to turn down a position that great scientists like Newton and Rutherford had occupied over the last 350 years, and it seemed a new and different sort of challenge. So in the end, I warned the vice presidents of my shortcomings and said if they still wanted me, I'd give it my best shot. They and the governing body (the council) chose to ignore my warnings and put me up for what a friend described as their usual North Korean style election, with only one name on the ballot.

Of course, nobody had taken the slightest interest in honouring me in these ways during the many years after the key breakthroughs with the ribosome; without my having won the Nobel lottery, none of them would have given me a second thought – if they had thought of me at all. So they were really all rewards for getting an award, and it reminded me of the line from Matthew 13:12: 'For whosoever hath, to him shall be given, and he shall have more abundance: but whosoever hath not, from him shall be taken away even that he hath.'

Tom Steitz too has found his life full of invitations and distractions. Among other things, he had a building named after him in his alma mater in Wisconsin. One year, he made four trips to China and seemed completely exhausted by his travels. I told him he could always say no. His lab continues to churn out high-profile papers in the general area of the central dogma, including many papers on the ribosome, some of which helped to settle the controversy with Ada over antibiotics unambiguously in his favour.

As the only woman alive who has won a Nobel Prize in Chemistry, Ada is in huge demand as a speaker and spends much of her time travelling around the globe. She is the recipient of a large number of honours and awards, including honorary degrees from both Oxford and Cambridge. On the one occasion that I visited the Weizmann Institute, I found an entire wall in her office covered with honorary degrees and awards. Although I had carefully timed my visit for dates she would be in town, she couldn't join me for either of the two dinners during my visit, she said, because her life was too hectic; she was just getting ready to travel abroad.

One of Ada's courageous acts after the prize was to say that all Palestinian political prisoners should be released, which got her a lot of criticism from the right-wing nationalists in Israel. She had introduced me by email to her friend the Zionist and pacifist Uri Avnery, and as a result I get his perceptive and witty essays on Israeli politics about once a week. Knowing her sentiments, I had suggested to her that the two of us give some lectures together at universities in the West Bank and at Al-Quds University in East Jerusalem. Despite Ada's sympathy for their cause, the Palestinians vetoed the idea, saying they were boycotting Israeli academics because of the occupation. So I ended up going without her. Curiously, my trip was arranged by Joe Zaccai, a Jewish scientist who works in Grenoble and was giving a short course in Birzeit University near Ramallah, so their objection was apparently not to Jews but specifically to Israelis. My visit to Israel and the West Bank left me pessimistic that there would ever be a solution to the Israeli-Palestinian problem.

What of the ones who were left out of the Nobel? If my mentor Peter Moore felt a sense of disappointment at having been left out of most of the awards for the ribosome, he did not show it. Rather, he seemed gratified at having helped see the ribosome structure emerge at Yale. He wound down his lab at the age of seventy and decided to focus on the esoteric problem of diffuse scattering. These are the X-rays scattered in between Bragg reflections in a crystal and contain information about parts of the molecule that are moving. It is the kind of problem that is hard and unfashionable, and probably only a few people both care enough and have the skills to understand it. Peter has never been one to fight in the trenches over trendy competitive problems, and my guess is that a challenging problem in which he will be left in peace to use his intellectual abilities suits him perfectly.

As someone whose lab had given us initial glimpses of many of the states of the ribosome, Joachim Frank, too, must have felt disappointed, although he graciously congratulated me that October.

He only had to wait a few more years. Once it became apparent that electron microscopy could yield the kind of resolution that could produce atomic structures, it was no longer blobology but had entered the realm of structural chemistry. In the autumn of 2017, as I was writing this, the Nobel Prize in Chemistry was awarded to Joachim Frank and Richard Henderson, along with Jacques Dubochet, who had figured out how to plunge biological samples into liquid ethane so they could be studied in their vitrified state at low temperatures.

None of the Nobel winners heard from Harry right after the prize, but he kept plugging away at the ribosome as he had for the previous several decades. As someone interested in cars and motorcycles, the engine aspect of the ribosome had always fascinated him, and a lot of his subsequent work involved trying to understand how the ribosome moves along the mRNA. His personal life had also taken a happy turn. Around the time of the breakthroughs in the atomic structures, he and his then graduate student Laura Lancaster began a relationship. They were married a few years later and continue to work together on their research. Harry's many admirers were outraged at his being left out of the prize and were determined to make amends. In 2016, he was awarded the Breakthrough Prize for his work. With eight times the cash value of a shared Nobel for the ribosome, he could certainly laugh all the way to the bank – or to a Ferrari dealership.

Marat and Gulnara returned to Strasbourg where they have been active in the ribosome field ever since. By producing the first high-resolution structure of an entire eukaryotic ribosome, Marat proved beyond all doubt that he was not a camel. A few years after the Nobel, he and Gulnara shared the Gregori Aminoff Prize for crystallography from the Swedish Academy of Sciences with Harry. Oddly, Jamie Cate, who had not only brought crystallographic expertise to the Santa Cruz group but whose use of osmium hexammine with the group I intron structure had been the key to obtaining phases for the ribosome structures, went unrecognized. It is a disgrace that he has still not been elected to the US National

Academy of Sciences, while those of us who benefitted from his idea have gone on to greater recognition.

Nenad Ban, now at the ETH in Zürich, and Poul Nissen, who returned to Århus, went on to have very successful careers and are among the leading structural biologists of their generation. Poul went into an entirely new area of how ions are pumped across cell membranes, but Nenad has continued to do some of the most important structural work being done on ribosomes and, like Jamie, is one of the people who gives my lab nightmares of being scooped.

Of the original 30S group in my own lab, Bil is at Caltech, Ditlev returned to Århus, and Andrew returned to the LMB after a long postdoc at UCSF. They are all very successful in their own research. Rob went on to get a PhD and then got a job in industry. James Ogle quit research to go into intellectual property law. After having worked on how the ribosome helps tRNA recognize the right codon, perhaps he felt anything else might be anticlimactic. He is also a gifted violinist and spends some of his time pursuing music on the side. Brian hated the idea of spending all his time teaching, managing people, and writing grant applications. After a long stint in industry for the antibiotic company that Tom and Peter founded (where ironically he overlapped and became friends with Francois Franceschi, Ada's former colleague), he now has a research position at the University of Colorado in Denver. He recently spent a sabbatical year in my lab at the LMB to learn electron microscopy, and although it reminded us of bygone days, we were both a lot older and it wasn't quite the same.

As for those of us who led the original ribosome structure groups, while many of us continued to be productive, we were no longer breaking really new ground. At best we were doing important work, but it was more of the same, which is fairly typical of people late in their career. Occasionally some do embark on something entirely new after the Nobel Prize, but frequently it gives them delusions of genius, and they overreach by quixotically tackling impossible problems for which they have no particular expertise. The very few people to have done new and fundamentally important work after the Nobel (and even more rarely won a second prize) did the work for their first one while they were still young enough to develop a completely new career.

The race for the ribosome structure raises general questions about competition and collaboration. As I have mentioned before, collaboration works best when people know each other well and like working together, or when they bring complementary expertise to bear on a problem that no one group can tackle alone. They are also essential for very large-scale projects, like the human genome project or the search for the Higgs boson, that can involve hundreds of people. These days there is a great deal of ideological fervour about the value of collaborations, but the truth is that scientists will collaborate or compete depending on what is in their self-interest. It is not a general rule that collaboration is always good and competition is always bad. Collaborations can get bogged down in the inertia and overhead of dealing with multiple people and labs and their associated bureaucracy. On the other hand, science is a marketplace of ideas, so just as in business, competition can spur people to think and work harder, weed out bad ideas and dead ends, and accelerate the pace of science. It can thus be good for science even if it isn't so great for scientists. Unlike in sport, the distinction between competition and collaboration is not so clear cut in science: even when scientists are competing, they are actually using one another's advances to make progress and are thus collaborating, albeit involuntarily.

What is also striking is that after a long struggle, several groups would make progress on the ribosome nearly simultaneously. This happens all the time in science and mathematics, even for discoveries we think are particularly great and profound. After centuries, calculus was simultaneously invented by Newton and Leibniz. The discovery of evolution by natural selection by Darwin and Wallace is another such example. Or two different formulations of quantum mechanics by Schrödinger and Heisenberg. Science never emerges from a vacuum. Rather, advances are made when certain ideas are in the air, and the state of understanding in a field and developments in technology reach a stage where those ideas can be pursued. When that happens, one or more people happen to see the next possible advance a little before everyone else. In the case of the ribosome, the development of synchrotrons, modern X-ray detectors, anomalous scattering, powerful computers and graphics, and cheap and abundant disk space were all essential for success, yet none of them was invented with the ribosome in mind.

So I don't subscribe to the heroic narrative of science. Rather, some of us are fortunate enough to be the agents of important discoveries that would have been made anyway, sometimes not even that much later. But this cold analytical view does not sit well with our emotional selves. We humans tend to personify everything we touch. We give names to theories and theorems, discoveries, laboratories, even pieces of apparatus. Science becomes a play, with heroes and villains. So even if discoveries are inevitable, we recognize that it is individuals who make them happen, and we like to honour those who took that first leap into the unknown, to go just beyond what was thought to be possible. And when someone like Newton or Einstein sees much further than others, or Watson and Crick synthesize in one stroke the essential features of DNA that might have dribbled out in pieces, we tend to immortalize them.

Looking back, it still astonishes me that my career worked out at all after so many false starts and dead ends. My beginnings were not promising. There were so many times when I could have fallen off the edge and disappeared from the world of science, a fate I only avoided by changing tack or starting over again. It was also a matter of luck that just when I needed them, smart people with just the right mixture of skills and attitudes joined the lab, and various friends and colleagues helped me in critical ways along the way. The story of the ribosome had its own drama, and whether we were mere agents of discovery or not, it was exciting to be there when it happened.

ACKNOWLEDGEMENTS

THIS BOOK WOULD NOT HAVE seen the light of day without two people: my agent John Brockman, who was enthusiastic about it from the time we first spoke, and Alex Gann who over many years kept encouraging me to write it. I am grateful to my editors, T. J. Kelleher and Eric Henney at Basic Books and Sam Carter at Oneworld, who took a chance on me even though I had never written a book before, patiently put up with my early drafts, and steered me in the right direction with their thorough and thoughtful advice.

I thank the many people who took the trouble to read early versions of the manuscript and give me useful feedback and comments and in many cases confirmed key events: Juliet Carter, Claire Craig, Mark Donnelly, Alex Gann, Steve Harrison, Graeme Mitchison, Peter Moore, Carol Robinson, Peter Rosenthal, Song Tan, and Steve White, and among my own former lab members, Rebecca Voorhees, Lori Passmore, Brian Wimberly, and Andrew Carter. I am especially grateful to Jennifer Doudna for not only reading the manuscript but writing a foreword for it. I thank Paul Margiotta for help with figures 2.1–2.5, 3.1, 3.2, 3.4, 14.2, 14.5, 17.1, and 17.2.

Brigitte Wittmann-Liebold and the late Volker Erdmann filled me in on parts of the early crystallization efforts in Berlin. Joel Sussman, Håkon Hope, and Leemor Joshua-Tor described the early days of cryocrystallography at the Weizmann Institute. Maria Garber gave me a detailed account of the early days of crystallization in Pushchino. Many others, as described throughout the book, clarified or confirmed various facts for me. I am grateful to them all.

Much of science is a collaborative enterprise, and my contribution to the story in this book would not have happened without the many talented and dedicated young scientists who joined my lab. In addition, there were many other students and postdocs who worked in my lab who are not part of this story but who were equally important to my work and career over the years.

Finally, and most importantly, I am grateful to my wife, Vera Rosenberry, who has been such a wonderful companion and friend for several decades. The ups and downs of science would have been difficult without the balance she gave to my life. Many times, she has uprooted herself cheerfully to move with me because of my career aspirations, first all over the United States and eventually to England.

NOTES AND SUGGESTED READING

THROUGHOUT THE BOOK, I HAVE used first names for people whom I know quite well personally and last names otherwise, except when introducing them or, in some cases, reintroducing them after a long gap. The main exceptions are in chapters 3 and 4, where the few people I know would have been surrounded by a sea of last names and caused some awkward juxtapositions.

This book is a memoir. It is a personal account of my involvement with the determination of the structure of the ribosome. Many aspects of it have to do with a race to be the first to determine an atomic structure of the ribosome. But each person who runs a race will experience it in his or her own way. So the book is not about the race for the ribosome but about *my* race. It describes my recollections of my impressions as the ribosome story unfolded while I was in the middle of it. It is not intended to be a history of the field and is certainly not a scholarly treatise. Much of it is a first-hand account. Although memory is notoriously fallible, I have been helped by an extensive email record of correspondence from the mid 1990s onwards. Many of the episodes describe public events. These and others have in most cases been confirmed by the people involved, especially when they have been directly attributed as sources in the text. The following are a few notes on parts of the story that I learned from others. I also include a few key papers and some suggestions for additional reading.

CHAPTER 2

One of the most readable, yet thorough accounts of how the information in DNA is coded to make proteins is *Life's Greatest Secret: The Race to Crack the Genetic Code* by Matthew Cobb (London: Hachette, 2015).

Sydney Brenner's comments on the ribosome are as quoted in *The Nematode Caenorhabditis elegans*, edited by W. B. Wood and the community of *C. elegans* researchers (Cold Spring Harbor, NY: Cold Spring Harbor Laboratory Press, 1988); and in F. H. C. Crick and S. Brenner, *Report to the Medical Research Council: On the Work of the Division of Molecular Genetics, now the Division of Cell Biology, from 1961–1971* (Cambridge, UK: MRC Lab of Molecular Biology, 1971).

The article in *Scientific American* that motivated me to write to Don Engelman was 'Neutron-scattering Studies of the Ribosome' by Donald M. Engelman and Peter B. Moore, *Scientific American* 235 (October 1976): 44–56.

CHAPTER 3

The discovery of molecules in the eighteenth and nineteenth centuries is a thrilling story told in *Chasing the Molecule* by John Buckingham (Phoenix Mill, UK: Sutton Publishing, 2004).

Lawrence Bragg himself wrote an excellent popular account of crystallography, 'X-ray Crystallography,' *Scientific American* 219 (July 1968): 58–74.

Henry Armstrong's scathing letter was published in the October 1, 1927, issue of *Nature*, on page 478.

The accounts of J. D. Bernal and Dorothy Hodgkin come from two excellent biographies: Andrew Brown, *J. D. Bernal: The Sage of Science* (Oxford, UK: Oxford University Press, 2005); and Georgina Ferry, *Dorothy Hodgkin: A Life* (Cold Spring Harbor, NY: Cold Spring Harbor Laboratory Press, 1998).

Two very readable articles about overcoming the phase problem for protein crystallography are by Max Perutz, 'The Hemoglobin Molecule,' *Scientific American* 211 (November 1964): 64–79; and in www.nobelprize.org/nobel_prizes/chemistry/laureates/1962/perspectives.html.

CHAPTER 4

The problem of crystallizing membrane proteins was overcome by Hartmut Michel, who was able to use special detergents to make them soluble and crystallize them. He went on to share the Nobel Prize in 1988 with Hans Deisenhofer and Robert Huber for the structure of the first membrane protein.

Nigel Unwin's early efforts were recounted to me personally by him. He is also the person who told me that Byers visited the LMB.

My account of the origins of ribosome crystallography in Berlin in this chapter was the result of a series of email exchanges with Brigitte Wittmann-Liebold and Volker Erdmann. I also had an extensive conversation with Erdmann that I wrote up as notes that he confirmed as accurate. Finally, I also talked with Knud Nierhaus, a leading ribosome biochemist in Wittmann's institute, over several years before he died a few years ago.

The Paradies episode was brought to light by Wayne Hendrickson and others in a letter, 'True identity of a diffraction pattern attributed to valyl tRNA,' *Nature* 303 (May 19, 1983): 195. Paradies's response was published on the following page, and the episode was discussed at length on page 197, including his departure from King's College and eventually the Free University in Berlin. In a recent email, Wayne Hendrickson completely stands by his analysis.

Bob Fletterick told me by email about his almost going to Wittmann's institute to work on the ribosome, and that he learned from Ada Yonath herself about the Humboldt fellowship being transferred to her.

Some facts about Ada Yonath's early life can be found in her autobiographical essay: www.nobelprize.org/nobel_prizes/chemistry/laureates/2009/yonath-bio.html.

The history of the Russian effort in Pushchino to crystallize the ribosome was shared with me by Maria Garber over a series of emails. Additional information was provided by Marat Yusupov and Alex Spirin.

CHAPTER 6

Details of Harry Noller's early career can be found in his very readable and entertaining autobiographical essay, 'By Ribosome Possessed,' *Journal of Biological Chemistry* 288 (2013): 24872–24885; also available online at www.jbc.org/content/288/34/24872.short.

CHAPTER 7

The description in this chapter of the origins of cryocrystallography is a result of extensive email correspondence with Håkon Hope and Joel Sussman. The final account was confirmed by them as well as by Leemor Joshua-Tor.

CHAPTER 8

In the chapter from Ada's group for the Victoria symposium (F. Schluenzen, H. A. S. Hansen, J. Thygesen, W. S. Bennett, N. Volkmann, I. Levin, J. Harms, H. Bartels, A. Zaytzev-Bashan, Z. Berkovitch-Yellin, I. Sagi, F. Franceschi, S. Krumbholz, M. Geva, S. Weinstein, I. Agmon, N. Boddeker, S. Morlang, R. Sharon, A. Dribin, E. Maltz,

M. Peretz, V. Weinrich, and A. Yonath, 'A Milestone in Ribosomal Crystallography: The Construction of Preliminary Electron Density Maps at Intermediate Resolution,' *Biochemistry and Cell Biology* 73 [1995]: 739–749), the 30S crystals were described as having a P42(1)2 symmetry instead of the actual P4(1)2(1)2 symmetry that both we and Ada's group established later. The difference is analogous to four molecules arranged along the corners of a square table versus four molecules arranged along a spiral staircase.

Regarding the 50S subunit, when the Yale group published their breakthrough paper three years later (N. Ban, B. Freeborn, P. Nissen, P. Penczek, R. A. Grassucci, R. Sweet, J. Frank, P. B. Moore, and T. A. Steitz. 'A 9 Å Resolution X-ray Crystallographic Map of the Large Ribosomal Subunit,' *Cell* 93 [1998]: 1105–1115), they stated, 'In contrast, our all-X-ray map does not resemble the previously published X-ray electron density map of the H. marismortui large ribosomal subunit, which had a nominal resolution at 7 Å, nor is the packing of subunits in the unit cell described here the same as that deduced in that study from a solvent flattened map (Schluenzen et al., 1995).'

CHAPTER 15

Crick's view on the Nobel Prize and its being a lottery is well articulated in this video interview: www.webofstories.com/play/francis.crick/75.

CHAPTER 16

'When you're Avis you have to try harder' refers to a famous ad campaign by the car rental company Avis in the 1960s and 1970s, when they were trying to catch up with Hertz. An account of it can be found in www.adweek.com/creativity/how-avis-brilliantly-pioneered-underdog-advertising-with-we-try-harder/.

CHAPTER 21

Peter Collins told me that the relevance of Matthew 13:12 to credit in science is so well known that it is the subject of an article by R. K. Merton, 'The Matthew Effect in Science: The Reward and Communication Systems of Science Are Considered,' *Science* 159 (1968): 56–63.

ABOUT THE AUTHOR

VENKI RAMAKRISHNAN is a structural biologist who shared the 2009 Nobel Prize in Chemistry for his work on the structure and function of the ribosome. He was knighted in 2012, and in 2015 was elected as President of the Royal Society. He works at the MRC Laboratory of Molecular Biology in Cambridge, England.